Pre-Crisis Planning, Communication, and Management

PETER LANG
New York • Washington, D.C./Baltimore • Bern
Frankfurt • Berlin • Brussels • Vienna • Oxford

Pre-Crisis Planning, Communication, and Management

Preparing for the Inevitable

EDITED BY
Bolanle A. Olaniran, David E. Williams,
and W. Timothy Coombs

PETER LANG
New York • Washington, D.C./Baltimore • Bern
Frankfurt • Berlin • Brussels • Vienna • Oxford

Library of Congress Cataloging-in-Publication Data

Pre-crisis planning, communication, and management: preparing
for the inevitable / edited by Bolanle A. Olaniran, David E. Williams,
W. Timothy Coombs.
p. cm.
Includes bibliographical references and index.
1. Emergency management. 2. Issues management.
I. Olaniran, Bolanle Abodunrin. II. Williams, David E.
III. Coombs, W. Timothy.
HD49.P74 658.4'056—dc23 2011042352
ISBN 978-1-4331-1135-8 (hardcover)
ISBN 978-1-4331-1134-1 (paperback)
ISBN 978-1-4539-0565-4 (e-book)

Bibliographic information published by **Die Deutsche Nationalbibliothek**.
Die Deutsche Nationalbibliothek lists this publication in the "Deutsche
Nationalbibliografie"; detailed bibliographic data is available
on the Internet at http://dnb.d-nb.de/.

The paper in this book meets the guidelines for permanence and durability
of the Committee on Production Guidelines for Book Longevity
of the Council of Library Resources.

29 Broadway, 18th floor, New York, NY 10006
www.peterlang.com

Printed in the United States of America

Contents

Section 2: Integrating Risk Communication into Pre-crisis Planning

Section 3: Managing Diversity in Crisis

Foreword

Robert S. Littlefield

When the National Weather Service broke the news in January 2011 that the Fargo, North Dakota, and Moorhead, Minnesota, communities would experience yet again record-level flooding of the Red River of the North and overland flooding from tributaries flowing into the Red River (National Weather Service, 2011), the populations braced themselves for the worst. Fourteen years earlier, Fargo-Moorhead residents witnessed in 1997 the impact of the flooding in Grand Forks, a neighboring city 70 miles to the north. When the Red River rose above the dikes on April 18, 1997, Grand Forks' infrastructure disintegrated, fire destroyed 11 downtown buildings (AP, 2002), and 46,000 people (90% of the population) were evacuated to the Grand Forks Air Force Base and surrounding communities (Draves, 1999) until the water receded and the restoration of the infrastructure—estimated at nearly $2 billion—commenced (Draves, 2002a).

While the city of Grand Forks recovered, physical and emotional scars remained for many Grand Forks residents (Draves, 2002b) as well as for other North Dakota and Minnesota residents with friends or relatives who experienced the disaster in Grand Forks. Within their own city limits, Fargo-Moorhead residents knew the devastating effects of flooding, experiencing a close call on March 31, 2009, when even their best efforts resulted in a near-catastrophic breach and flooding of several buildings on the Oak Grove Lutheran School campus in north Fargo (Smith, 2009). The floods of 1997 and 2009 were just two of many floods affecting the Red River Valley over the past 30 years (Hasbargen, 2011). Because of the propensity for flooding, especially during the recent wet year cycle when flood levels in 2011 were predicted to top the 2009 records (Marin, 2011), Fargo and Moorhead city leaders quickly moved forward with their planning and subsequently put their plans into action.

As Olaniran and Williams suggest in this volume, crisis communication scholars acknowledge the three-phase crisis model described by Seeger, Sellnow, and Ulmer (2003) as a straightforward and easily discernable way to understand what happens as individuals, organizations, or communities pass through the phases of a crisis. Using this model, observers move from

(1) the pre-crisis phase where signs appear signaling the potential for a crisis to occur, (2) to a full-blown crisis phase where chaos results due to the disequilibrium brought on by the crisis itself, (3) and to the post-crisis phase where order is restored, lessons are learned, and steps are taken to avoid experiencing a similar situation in the future. This model is cyclical in that as it moves through the post-crisis phase, the next pre-crisis phase is beginning.

At the end of this model, the movement from post-crisis to pre-crisis phases provides the calculus for those seeking to understand why some crisis managers are able to avoid catastrophe while others are not. The identification of crisis types (Coombs, 2007; Crandall, Parnell, & Spillan, 2010) suggests that even though particular crises may have never been previously experienced first-hand, all crises fall into types that can be analyzed and from which learning may occur. If experience must be the teacher, then those crisis managers and communicators who passed through the three phases of a crisis should have insight into those practices that succeeded or failed. For those without the actual experience of moving through the three crisis phases, being attentive to what others have learned should provide direction and lead to more successful management strategies in the future.

The example of the Fargo-Moorhead flood of 2011 brings to mind important dimensions of crisis communication that occur during the pre-crisis phase: anticipation of the crisis, effective communication, and management of strategies to prevent or mitigate the crisis. As this anticipatory perspective is unpacked, its complexity is revealed. First, anticipation implies vigilance or mindfulness (Langer, 1989) that the potential for crisis is ever present. Langer characterizes the mindful state of being as "(1) creation of new categories; (2) openness to new information; and (3) awareness of more than one perspective" (p. 62). While it may not be possible to anticipate every particular crisis, there are specific indicators that signal the chance for a crisis to occur. Thus, from a mindful perspective, one should be aware of new categories, be open to new information, and be cognizant of more than one perspective in anticipation of a crisis without ever having experienced it first-hand based upon warnings about how the indicators will manifest themselves. In contrast, anticipation might be prompted based upon prior experience with a particular crisis. In this latter context, by managing a crisis situation from pre-crisis through the post-crisis phase, the actual experience itself might serve as the motivation for vigilance. Watching for signs becomes part of the pre-crisis process. Thus, in the case of the Fargo and Moorhead flood crisis, the experience of the community leaders in managing previous floods led them to be vigilant as they anticipated the predictions of the National Weather Service and began to utilize some of their own personal observation techniques to determine their perception of the flood risk.

The vigilance of the leaders also prompted communication and strategies requiring future action. Using the anticipatory framework, community leaders know that not only will the content of their messages be interpreted, but also that the means by which those messages are transmitted to community stakeholders will become important. As the threat of crisis becomes more imminent, spokespersons must begin crafting messages that prepare the community residents for what lies ahead. Not only must they communicate information about the nature and potential of the flood risk, but they also must provide opportunities for public efficacy as residents and other stakeholders seek ways to protect themselves and their property from potential damage. In the case of the Fargo flood, the leaders used various means of communicating with community members including public forums, public meetings, press conferences, media (newspaper, radio, and television), social media (Internet, Facebook, Twitter, and blogs), telephone (including reverse 911), and face-to-face interaction. While the means and methods of communication used in 2011 were more technologically advanced than those used in 1997 (Morse, 2011), effective communication between city leaders, National Guard troops, city crews, community volunteers, and various civic organizations was essential with the river stretching the length of Fargo-Moorhead's core. In addition to communication with the formal stakeholders, over time leaders recognized the need to include the different publics in the crisis response dialogue. The leaders reached out through neighborhood meetings and through social service organizations to provide for the security of those individuals considered vulnerable due to age, health, socio-economic status, or education level.

Finally, the collective leadership of the community working together to manage the strategies mitigating the damage of the flood contributed to the success of the effort. Considerable learning occurred during the years 1997 to 2011—such that when the planning began in 2011, the leaders included particular strategies that were effective and eliminated those that were not. For example, instead of sandbagging the entire riverfront, some clay levees were built to secure entire neighborhoods. To facilitate the fast transfer of sandbags from their warm storage facilities used to prevent them from freezing, law enforcement officers were used to control traffic on North-South thoroughfares during particular times of the day. Similarly, improved technology enhanced efficiency. For example, spider-like sandbag-filling machines were used in 2011 to channel the sand into 10 sandbags simultaneously, replacing the one-shovel-to-one-bag system used in 1997 ('Sandbag Central' opens, 2011). Quite clearly, because the community leaders were vigilant in anticipation of a potential flood crisis, utilized varied and effective means of communication, and managed strategies that would mitigate or prevent damage from the flood, Fargo and Moorhead leaders helped their communities avoid the devastation experienced by other towns and cities in the region.

Particularly since 1997, the mayor and community crisis teams in Fargo and Moorhead have benefited from collectively experiencing the flood cycle repeatedly. On several occasions, the crisis cycle has moved from pre-crisis warnings of potential flooding, to the real crisis of rising river levels and major flooding, to the use of dikes and other means to prevent damage to private and public property, and to post-crisis learning and assessment about what worked and what did not. Because the individuals on the leadership teams have remained fairly constant over the years (e.g., the current Fargo mayor, the Fargo city engineer, and the Fargo city manager have served simultaneously in official capacities since 1997), their familiarity with each other's strengths and limitations, as well as their collective memory and experiences in managing the flood crises, has enabled them to take the necessary steps to avoid major catastrophe.

Not everyone has the opportunity to learn first-hand how to communicate and manage a crisis situation. It is for these individuals who must learn through the experiences of others that this volume may be particularly useful. While phases of a crisis function in a cycle and provide an opportunity to examine communication from different vantage points, this book concentrates on the pre-crisis stage because it is at this point that crisis management has the opportunity to have its greatest impact mitigating the effects of the crisis.

In early studies of crisis, scholars analyzed the communication and actions of crisis managers and executives as they attempted to mitigate the damage and rebuild their images as well as the public's trust in their company or product. This post-crisis assessment relied on the crisis itself as the catalyst for the research, providing an indication of how the individual, organization, or community responded under crisis. In contrast, by addressing the anticipatory dimension of crisis management, communicators and managers have a greater opportunity to prepare for and mitigate the effects of a crisis or natural disaster.

In 2011, the Fargo and Moorhead communities braced themselves for a flood that was predicted to surpass the highest water levels they had ever experienced (Valley Flood Watch, 2011). To avoid having to produce millions of sandbags under the pressure of rising river levels—as was the case in March of 2009—volunteers from the two communities were called into action on February 14, 2011, in order to fill the three million sandbags needed to protect the community. Residents were urged to prepare their own homes and told how to safeguard their possessions in the event that the waters breached the dikes and levees. High school and college students, as well as prisoners from the county jail, were drafted into service, and college campuses planned for cancelled classes, alternative delivery systems, and safety-related issues. News broadcasters reported on the flood fight nightly, and public meetings began. While regular warnings came from the National Weather Service, the community seemed to pay less attention, preferring to take more seriously the public pronouncements of the mayor, likened by some as a religious figure similar to Moses who was reported to have parted the waters to protect the people of Israel. The mayor's experience as a successful crisis manager of previous floods gave him ascribed status; and although the flood of 2011 did not materialize to the extent that was predicted, the anticipatory perspective taken by those leading the flood fight resulted in a community that was well prepared for what would come. As the water rose, the planning paid off. The water was contained, and the flood crisis of 2011 was averted.

The impact of all this seems clear: By focusing awareness on how an anticipatory perspective to crisis can be beneficial, this book can be a valuable resource for crisis communicators and managers as they move through the phases of a crisis.

References

Associated Press. (2002, April 13). City proud of recovery from 1997 flood. USA Today. Retrieved from www.usatoday.com

Coombs, W. T. (2007). *Ongoing crisis communication: Planning, managing, and responding* (2nd ed.). Thousand Oaks, CA: Sage.

Crandall, W. R., Parnell, J. A., & Spillan, J. E. (2010). *Crisis management in the new strategy landscape.* Thousand Oaks, CA: Sage.

Draves, A. (1999). Grand Forks flood and recovery. Retrieved from www.draves.com/gf/gindex.htm

Draves, A. (2002a). East Grand Forks five years later. Retrieved from www.draves.com/gf/egf5yrs.htm

Draves, A. (2002b). The 1997 flood in Grand Forks North Dakota. Retrieved from www.draves.com/gf/

Hasbargen, K. C. (2011). The Red River spring flood of 2009: Revealing leadership and spokesperson credibility through the discourse of renewal (Unpublished doctoral dissertation). North Dakota State University, Fargo, North Dakota.

Langer, E. J. (1989). *Mindfulness.* Cambridge, MA: Da Capo Press.

Marin, L. E. (2011, January 23). Fargo Flood 2011…Another spring flood ahead? Blog at WordPress.com. Retrieved from http://bigfishservicecompany.wordpress.com/2011/01/23

Morse, J. (2011, February 8). Civil society and social media. Retrieved from http://blogs.america.gov

National Weather Service. (2011, January 18). Preliminary 2011 spring flood outlook. Grand Forks, ND Weather Forecast Office, Grand Forks, ND. Retrieved from www.crh.noaa.gov/news

'Sandbag Central' opens in Fargo. (2011, February 14). MPR News. Retrieved from http://minnesota.publicradio.org

Seeger, M. W., Sellnow, T. L., & Ulmer, R. R. (2003). *Communication and organizational crisis.* Westport, CT: Praeger.

Smith, K. (2009, March 31). Oak Grove School begins cleanup after breach. INFORUM, Forum Communications Co. Retrieved from www.inforum.com

Valley Flood Watch. (2011, January 18). Preliminary 2011 flood outlook shows major flooding possible for the Valley. Go Radio Broadcasting, LLC. Retrieved from http://valleyfloodwatch.com

Introduction

Bolanle A. Olaniran & David E. Williams

The evolution of crisis communication research is vaguely ironic in that, after a 30-year history, the focus has progressed back to the initial stage of the management process. There are a few ways to view the life span of an organizational crisis. Fink (1986) led the way for most crisis experts when he wrote of the prodromal acute, chronic, and resolution stages of crisis. These metaphorical descriptions translate to signs of a potential crisis, triggering event, response action, and termination of the crisis. Ian Mitroff (1994) offered another model, which Coombs (2002) suggests has strong similarities to that authored by Fink. Mitroff's five-stage model includes signal detection, probing and prevention, damage containment, recovery, and learning. The initial two stages are areas of concern for this volume, as they draw attention to warning signs of crises and risk identification and reduction. The final three stages address the issues of crisis response and post-crisis responsibilities that direct the organization to review its response procedures, determine what did and did not work, and record the actions for future use.

Probably the most commonly considered model in professional and trade literature is the three-stage model, which Combs (2007) suggests cannot be attributed to a single source. However, one such work with a good description of the approach can be found in Seeger, Sellnow, & Ulmer (2003). The extensive use of this three-stage model likely stems from its clarity and simplicity. The pre-crisis, response, and post-crisis view is all-encompassing and allows for clear, albeit general, distinctions between crisis management needs, actions, and responsibilities.

The focus of this book is clearly on the pre-crisis stage. The irony of this focus comes in light of this new focus on pre-crisis planning emerging after a 30-year history of crisis research. We consider the pioneering works in crisis management and communication to be the works of Fink (1986), Lerbinger (1986), and Barton (1993). While other works can be found, even works that precede these, the work of these authors helped foster greater recognition of the importance of crisis management efforts to both practitioner and scholarly audiences. Each of these works offered an all-encompassing view of crisis management with full recognition of all three stages of the crisis model. In the time since these works were published, scholarly attention has traced the crisis life span in reverse. In general, much of the early attention to crisis management and communication focused on the latter response and post-crisis stages. While attention to pre-crisis issues has always been present, it is not until now that it has been a research focus instead of the groundwork for what was to come later.

Much of the early research on crisis communication addressed the success or failure of response efforts. Representative of this line of research are the works authored by Benoit & Lindsey (1987), Ice (1991), Schultz & Seeger (1991), Small (1991), Williams & Treadaway (1992), and Sellnow (1993). At least two prominent similarities can be noted in these representative works. First, they each focused the response efforts of a company facing crises. Rhetorical strategies, use of apologia specifically, scientific argument, and image reconstruction were all addressed in terms of crisis response strategies in these works. The authors analyzed the spoken and written communication from the organizations and their spokespersons, factored in the relevant obstacles and environmental concerns, and assessed the usefulness or effectiveness of the communication and response strategies.

The second similarity is common to many of the research efforts by this period. The research was driven by the crisis event. These authors analyzed crises responses as they related to the high profile *Exxon Valdez* oil spill, Tylenol tampering, and Chrysler bankruptcy case. The crisis and its significance were the driving forces behind the research and its importance. The response was important, but the research was generated because of the magnitude of the crisis event. A notable exception to the event-driven response research was the work on image restoration strategies (Benoit, 1997), which offered a theoretical view of how organizations construct responses to crises.

Growth in crisis management research drove scholarly attention backward in the three-stage model toward the response phase. More specifically, works began to emerge that recognized the important differences in crisis types. Again, not that the recognition was completely absent in previous literature, but it did grow into more of a focus.

Coombs (2002) and Seeger (2002) each penned articles representative of this new foci with their articles, which helped to further define the nature and distinctive qualities of a crisis. Research in this vein began to question what inherent differences exist in crisis types and how responses can be altered according to those differences. Coombs (2007) offers what he calls a “master list” of the various crisis types. These types include natural disasters, workplace violence, rumors, malevolence, challenges, technical-error accidents, technical-error product harm, human-error product harm, human-error accidents, and organizational misdeeds. Many authors would now add terrorism as a separate crisis type, while Coombs includes it as a malevolent act. While comprehensive, Coombs is not the only perspective on crisis types. For example, Crandall, Parnell, and Spillan (2010) suggest crisis sources are political-legal, economic forces, social forces, and technological forces.

The purpose of examining crisis types emerged early as seen in work by Coombs (1995). In this work Coombs aligned crisis types (as he characterized them at that time) with the variables of evidence, damage, victim status, and performance history to create a matrix outlining the prescribed crisis response strategy. This development in crisis research signaled an identification that one response strategy could not serve as a cure-all. The notion of tailoring responses to crisis type with consideration to other relevant factors has continued to grow. Recent scholarship has revealed a desire among scholars to identify industry-specific response strategies.

An early example of industry-specific crisis response was spurred on by the Columbine shooting. Seeger, Heyart, Barton, and Bultnyck (2001) examined the Michigan crisis response plan for school violence incidents. Other early industry-specific research foci included airline industry studies and natural disaster response analyses. More recent industry-specific research has been seen in oil industry crises (Maresh & Williams, 2010), racially oriented crises (Liu, 2010; Williams & Olaniran, 2002), and terrorism (Ulmer, Sellnow, & Seeger, 2007). As with the earlier post-crisis studies, these response and crisis type studies were driven by the crisis event. The prevalence of crises in a particular industry, as in oil and coal accidents, or the severity of damage, as in airline accidents and terrorists' threats and acts, establish the need to address the crises according to their similarities.

The evolution of crisis communication research has progressed (in reverse fashion) from post-crisis responses, to response and crisis types, to the focus of this work, pre-crisis management and communication. The notion of crisis preparation is certainly not new as evidenced by the early work of Mitroff (1986). The pioneers of crisis communication research addressed the need for pre-crisis planning through (what we refer to as) anticipation and the development of crisis communication plans as well as adequate preparations to allow for a proactive response to the crisis. However, a concerted focus on pre-crisis planning and anticipation through analysis and theory development has been absent until now. "The anticipatory model of crisis management is among the limited research in this area" (Coombs, 2010, p. 25).

This volume is the first book-length work devoted to the pre-crisis phase of crisis management and communication. As you will see, this new stage in the evolution of crisis research will be driven less than the previous two phases by the crisis event and more by the need to prevent the crisis and the theoretical insights that make anticipation possible.

The current volume does not view the three stages (i.e., pre-crisis, response, and post-crisis) as necessarily mutually exclusive but rather sees pre-crisis as the driving force or foundation for successful crisis management regardless of the stages (i.e., crisis response and post-crisis). From this perspective, both academic and practitioner audiences will be able to find something useful for their respective tasks. This approach targets communication, business/management, and public relations fields and disciplines. The book also offers both theoretical guidance and applied principles that form the basis for which crises can be analyzed, managed or implemented, while also focusing on how relations with stakeholders and or publics can be restored. Another important contribution of this volume is the usefulness and adaptability of principles that can be applied to any given crisis situation. The volume is organized into three sections: 1) Applying Theory to Pre-Crisis Planning, 2) Integrating Risk Communication into Pre-Crisis Planning, and 3) Managing Culture and Diversity in Crisis. This Introduction identifies research premises and foundations for the book.

The first section of the volume focuses on useful theoretical perspectives in pre-crisis efforts. Chapter 1 by Olaniran and Williams, "The Need for an Anticipatory Perspective in Crisis Communication," emphasizes and reinforces the need for pre-crisis planning, communication, and management as a foundation upon which subsequent chapters are based. The chapter offers recognition of the critical role that planning and anticipation play in successful efforts to respond to crises. This chapter highlights and explains the emergence of the anticipation focus and provides a review of the anticipatory model of crisis management (Olaniran & Williams, 2001; 2008), which is followed by a brief example that highlights the need for further insights into the anticipatory perspective of crisis management and communication.

Chapter 2 by Jaques, "Issue Management as a Strategic Aspect of Crisis Prevention Discipline," looks at the evolution of issue management in crisis as a way for organizations to participate and not simply respond to public in policies. In other words, issues management theory enables the practitioner and the organization to lead the search for troublesome issues before being obligated to address them.

Chapter 3 by Falkheimer and Heide, "Participatory Pre-crisis and Crisis Communication: A Conceptual Approach," argues the need for crisis communication to be flexible in the face of globalization of different communication media.

Chapter 4 by Bowen, "The Ethical Challenges of Pre-crisis Communication," attempts to answer a key question of what needs to be communicated and how it should be communicated prior to a crisis in order to make an organization ethical. The author argues that successful issues management should help to prevent a crisis, but it should also be carried out from a foundation of moral responsibility using both utilitarian and deontological analyses. The chapter discusses ethics of pre-crisis planning through the use of adapted steps in issue management that include ethical an decision-making framework.

Chapter 5 by Chandler, "Recognizing, Anticipating, and Preventing Ethical Misconduct Disasters," reviews the concepts of strategic ethical integrity and measures required for recognizing the warning signs of ethical crises, scandals, and ethical misconduct disasters. The chapter also alludes to the process and benefits of conducting an Ethical Conduct Audit©, which is presented along with guidelines for preventing ethical crises and ethical misconduct disasters.

The second section, Integrating Risk Communication into Pre-Crisis Planning, opens with Chapter 6 by Heath and Sultan, "Pre-crisis Management and Communication: Slippery Steps or Solid Footing?" The authors reinforce the foundation addressed in the Introduction and set the stage for the discussions in the subsequent chapters. Using different references and examples, the chapter identifies challenges faced by private sector organizations and government and non-profit organizations where executives fail their organization and their stakeholders/stakeseekers. Consequently, it reinforces the fact that organizations of all types need pre-crisis management and communication policies along with good implementation strategy.

Chapter 7 by Jin, Pang, and Cameron, "Pre-Crisis Threat Assessment: A Cognitive Appraisal Approach," examines a pre-crisis threat assessment model by introducing a cognitive appraisal approach to evaluate the situational demands and allocated organizational resources in order to enhance organizational preparedness and to address the pre-crisis issues more effectively.

Chapter 8 by Palenchar and Freberg, "Emergency Management Planning: Risk, Crises, Issues and Social Media," focuses on what needs to be communicated and how it should be communicated prior to a crisis. The authors argue in favor of an early forecasting system and a proactive public relations and risk communication program.

Chapter 9 by Low, Chung, and Pang, "Efficacy of Crisis Management Plans: A Revised Synthesis Model in Managing Crises" stresses the critical role crisis management plans (CMP) have in providing a platform to engage organizations in developing a systematic program in stakeholder mapping; risk and issues identification; delineating the roles and responsibilities of the crisis management teams; and developing contact lists for different stakeholders. Their work identifies two trigger points that call for the activation of the CMP when industry partners encounter a crisis.

In the third and final section, which focuses on managing culture and diversity in crisis, the section opens with Chapter 10 by Veil, Mitcham, and Sellnow. "Lost in Translation: How Emergency Managers Communicate Preparedness with Vulnerable Populations" discusses how vulnerable populations (e.g., low income and ethnic minority groups) face additional danger in crises due to socio-economic challenges lack of and access to communication media for evacuation orders and the like.

Chapter 11 by Olaniran and Williams, "A Multicultural Perspective on Crisis Management: Shell and the Ogoni People of Nigeria," focuses on the impacts of culture in crisis management. Specifically, the chapter addresses the need to manage crisis where two or more different cultures are in play in a given crisis. The chapter illustrates this point with the Nigerian Ogoni and Shell crisis.

Chapter 12 by Ni and Vardeman-Winter, "China: Growing Economy and Growing Crisis Potential," explores a cultural model of pre-crisis planning in the context of China. The chapter focuses on factors that may influence or predict effective pre-crisis planning. These factors were then discussed in the framework of general pre-crisis planning principles. The chapter concludes with suggestions on the methods for studying pre-crisis planning.

Chapter 13 by Waymer, "Crisis Management and Communication: Pre-crisis Preparation That Is Sensitive to Diverse Multicultural Populations," extends the logic of the anticipatory model of crisis management into areas of cultural diversity. Specifically, it establishes that due to the continual influx of ethnically and racially diverse populations, organizations must be vigilant in crisis preparation and in a manner that is sensitive to multicultural populations.

The section wraps up with Chapter 14 by Williams and Olaniran, "Racially Oriented Crisis: A Unique Challenge to Crises Preparation." The authors identify three qualities that make racially oriented crises unique and difficult to manage, including the fact that these crises are difficult to anticipate, damage targets the organizational or individual's reputation, and responses tend to require the crisis response team to choose among polarized strategies. The chapter stresses that crises involving race can range from a single spoken ethnic slur to a policy or practice that invokes depictions of a history of bias and prejudice. As a result, the authors suggest the need to focus on pre-crisis by putting a high priority on monitoring speeches of high-level executives in organizations or other prominent individuals (e.g., celebrities).

Each chapter thoroughly presents the foundation of pre-crisis planning, communication, and management. It is our hope that organizations, practitioners, and stakeholders find this volume a breath of fresh air and an approach that, when done right, will yield big dividends and consequently help them in accomplishing their goals of heading off crises, communicating crisis plan efforts, and effectively managing crises in a manner that includes and appeals to different publics.

References

Barton, L. (1993). *Crisis in organizations: Managing and communicating in the heat of chaos*. Cincinnati, OH: College Divisions South-Western.

Benoit, W. L. (1997). Image repair discourse and crisis communication. *Public Relations Review*, *23*(2), 177–186.

Benoit, W. L., & Lindsey, J. J. (1987). Argument strategies: Antidote to Tylenol's poisoned image. *Journal of the American Forensic Association*, *23*, 136–146.

Coombs, W. T. (1995). Choosing the right words: The development of guidelines for the selection of the appropriate crisis-response strategies. *Management Communication Quarterly*, *8*, 447–476.

———. (2002). Deep and surface threats: Conceptual and practical implications for "crisis" vs. "problem". *Public Relations Review*, *28*, 339–345.

———. (2010). Parameters for crisis communication. In W. T. Coombs & S. Holladay (Eds.), *The handbook of crisis communication*. Malden, MA: Wiley-Blackwell.

Crandall, W. R., Parnell, J. A,, & Spillan, J. E. (2010). *Crisis management in the new strategy landscape*. Los Angeles: Sage.

Fink, S. (1986). *Crisis management: Planning for the inevitable*. New York: AMACOM.

Ice, R. (1991). Corporate publics and rhetorical strategies: The case of Union Carbide's Bhopal crisis. *Management Communication Quarterly*, *4*, 341–362.

Lerbinger, O. (1986). *Managing corporate crises: Strategies for executives*. Boston, MA: Barrington.

Liu, B. F. (2010). Effective public relations in racially charged crises: Not black or white. In W. T. Coombs & S. J. Holladay (Eds.), *The handbook of crisis communication*. Malden, MA: Wiley-Blackwell.

Maresh, M., & Williams, D. E. (2010). Oil industry crisis communication. In W. T. Coombs & S. J. Holladay (Eds.), *The handbook of crisis communication*. Malden, MA: Wiley-Blackwell.

Mitroff, I. I. (1986). Prevention: Teaching corporate America to think about crisis prevention. *Journal of Business Studies*, *6*, 40–48.

———. (1994, Winter). Crisis management and environmentalism: A natural fit. *California Management Review*, *36*(2), 101–113.

Olaniran, B. A., & Williams, D. E. (2001). Anticipatory model of crisis management: A vigilant response to technological crisis. In R. Heath and G. Vasquez (Eds.), *Handbook of public relations* (pp. 487–500). Thousand Oaks, CA: Sage.

———, & ———. (2008). Applying anticipatory and relational perspectives to the Nigerian Delta Region oil crisis. *Public Relations Review, 34*, 57–59.

Schultz, P. D., & Seeger, M. W. (1991). Corporate centered apologia: Iacocca in defense of Chrysler. *Speaker and Gavel*, *28*, 50–60.

Seeger, M. W. (2002). Chaos and crisis: Propositions for a general theory of crisis communication. *Public Relations Review*, *28*, 329–337.

Seeger, M. W., Heyart, B., Barton, E. A., & Bultnyck, S. (2001). Crisis planning and crisis communication in the public schools: Assessing post-Columbine responses. *Communication Research Reports*, 375–383.

Seeger, M. W., Sellnow, T. L., & Ulmer, R. R. (2003). *Communication and organizational crisis*. Westport, CT.: Praeger.

Sellnow, T. L. (1993). Scientific argument in organizational crisis communication: The case of Exxon. *Argumentation and Advocacy*, *30*, 28–42.

Small, W. W. (1991). *Exxon Valdez*: How to spend billions and still get a black eye. *Public Relations Review*, *17*, 9–25.

Ulmer, R., Sellnow, T. L., & M. W. Seeger (2007). *Effective crisis communication: Moving from crisis to opportunity*. Thousand Oaks: Sage.

Williams, D. E., & Olaniran, B. A. (2002). Crisis communication in racial issues. *Journal of Applied Communication Research*, *30*(4), 293–313.

Williams, D. E., & Treadaway, G. (1992). Exxon and the *Valdez* accident: A failure in crisis communication. *Communication Studies*, *43*, 56–64.

Section 1

Applying Theory to Pre-crisis Planning

Chapter 1

The Need for an Anticipatory Perspective in Crisis Communication

Bolanle A. Olaniran & David E. Williams

This chapter sets the foundation for the book, as it introduces the anticipatory perspective. The anticipatory perspective is one of the earliest attempts at theory development designed specifically for the pre-crisis phase of crisis management. This chapter recalls the inception of the model and highlights its major components. A brief case study is offered to push the anticipatory perspective from theory into practice.

Introduction

The study and practice of crisis communication have evolved significantly over the last two decades. Aside from developments in the day-to-day practices of crisis management, there has been congruent growth in theoretical development, media attention, and public interest in crisis, crisis management, and, therefore, crisis communication. Within the realm of theoretical development in crisis management and communication is the recognition of the critical role that planning and anticipation play in successful efforts to respond to crises. This chapter lays the foundation for the importance of anticipation and pre-crisis planning that will be explored in chapters that follow. An explanation of the emergence of the anticipation focus and a review of the anticipatory model (Olaniran and Williams, 2001) are followed by a brief example that highlights the need for further insights into the anticipatory perspective of crisis management and communication.

The Inception of Anticipation

The need to adopt a proactive stance was one of the early admonitions for public relations practitioners working on crisis teams. Earlier work (e.g., Fink, 1986; Lerbinger, 1997) noted the importance of getting the organizational perspective to reporters in a timely manner in order to help shape media and public perception of the crisis. This effort involved the need to maintain healthy media relations and the ability to respond quickly following any type of crisis. There was less attention paid initially to types of crises or even particular strategies used in accordance with crisis situations. However, there are some notable exceptions, such as Perrow's (1884) work on "normal accidents." In this work, Perrow addressed issues related to crisis and risk with technology and technological failure. While not identifying other pre-crisis efforts as "anticipatory," crisis researchers did specify the need for crisis plans and pre-crisis planning. These researchers (e.g., Barton, 1993, 2001, 2007; Mitroff, 1986, 2003) have been at the forefront of setting the professional benchmarks for crisis preparedness and leadership. This body of work set the foundation for understanding the need and function of crisis planning for the inevitability of crises. Their writings and those of others detailed the pragmatic necessity to develop crisis teams and practice crisis scenarios regarding who would speak, what would be said, and what actions the organization would take. They also recognized the importance of crisis audits in order to determine the most likely vulnerabilities or sources of future crises. These pre-anticipatory orientation efforts were geared to supporting the implementation of a quick organizational response. A *proactive* response was one of the key buzz words of early crisis response research—one that is still applicable today. Benson's (1988) excellent work on the Tylenol response was one of many that addressed the success of Johnson & Johnson in managing the tampering crisis. Benson credited the *proactive stance* utilized by Johnson & Johnson as one of the key factors in their ability to survive, and even thrive, following the crisis.

Research momentum escalated through the late 1980s into the 1990s and was certainly spurred on by the events of 9/11. The 9/11 attacks helped draw attention to a shift in crisis management and communication that was already underway. This shift was a move beyond a response orientation to an anticipatory orientation. It would be overstating the case to say that the anticipatory orientation replaced or overshadowed the response orientation. However, the anticipatory orientation began to take on a level of coexistence with the response orientation. Crisis managers and researchers began to see the work of crisis anticipation and crisis response as equally important, deserving of equal time and effort in preparation and practice, and deserving of equal attention from scholars. Scholarly work on crisis response, led by the works of Coombs and Heath, became more precise (e.g., Coombs, 1995; Heath & Millar, 2004) while work on the anticipatory orientation also became more intricate (e.g., Olaniran & Williams, 2001; Pauchant & Mitroff, 1992).

The notion of pre-crisis planning transformed in the span of about a decade from a focus on media relationships, quick responses, and detailed planning of actions and communications to a focus on initiating a much more complex matrix involving greater research and decision-making considerations. Coombs (2007) does an excellent job of outlining many of the components of this *new* anticipatory orientation. Coombs and others have drawn attention to elements of crisis planning, which fit into what we refer to as the anticipatory orientation to crisis management. These include recognizing the importance of crisis history for an organization, managing reputation, assessing risk, recognizing crisis vulnerabilities, and understanding differences in crisis types.

The *new* anticipatory orientation toward crisis management moves the starting point for crisis evaluation. In the early days of crisis research, the starting point was largely viewed as the onset of the crisis. While suggestions and plans were offered for pre-crisis planning, much of the effort was focused on the start of the crisis, the initial response, and then whatever followed. Now, the understanding of and preparation for crisis has a significantly different starting point with a more complete understanding of the inevitability of crisis events. For example, the recognition of crisis history and organizational reputation alerts practitioners and researchers that public and stakeholder understanding and interpretation of a crisis event is partially shaped by what they already know of the organization. A poor history of crisis response will certainly taint views of that organization if they encounter another crisis. Public recognition of the *Exxon Valdez* grounding will still be invoked (two decades later) if Exxon has a similar tanker incident, particularly in Alaskan waters. We would argue that the crisis history can extend beyond prior actions of the organization. In certain cases industry crisis history can negatively affect an organization that meetswith crisis for the first time. Notably, over the last few years celebrities have been caught in the midst of crises stemming from their actions.

Broadcaster Don Imus, comedian Michael Richards, and sports reporter Kelly Tilghman were all involved in racially oriented crises (or ROCs, as will be discussed in a later chapter). With each subsequent incident, the public was reminded of previous inappropriate statements made by celebrities or media members. This history, although separate from each individual, wrapped each of them together in the public consciousness of racially insensitive statements. In essence, a celebrity who utters a racially insensitive comment takes the public memory back to other similar statements made in recent history. Therefore, the crisis has a starting point before the most recent racial slur.

Examining crisis vulnerabilities develops the early recognition by Fink (1986) that particular industries will be susceptible to particular crises. Combining knowledge of vulnerabilities with crisis types that are more typical for certain organizations or industries allows practitioners to more proactively determine where preparatory and prevention efforts should be devoted. Combining research foci in crisis management/communication and risk management/communication (e.g., Williams & Olaniran, 1998; Seeger, Sellnow, & Ulmer, 2001) is an example of the development of Fink's original observation.

Research over the past several years has provided practitioners with insights into how to identify the areas where they may be more susceptible along with specific information on how to best manage different crisis types. For example, an organization may be equally susceptible to crises involving natural disasters and organizational misdeeds. A company in California designed to handle hazardous waste cleanup could determine it needs to prepare for an earthquake that could damage waste containers, thus creating a hazardous waste spill. Additionally, the organization might see worker or management shortcuts that cause environmental threats as a possible crisis risk. Each of these crisis types (natural disaster and organizational misdeed) are plausible and must be prepared for both in terms of management actions and crisis communication strategies to be employed. However, the same preparations and communication strategies would not function optimally for both crisis types. The *new* anticipatory orientation assists public relations professionals and all others involved in crisis and emergency response with the recognition that crisis preparation cannot be performed with one management action or one response strategy. Instead, it is an ongoing process of monitoring and reassessing crisis risks. The practitioner, now more than ever, is asked to look into organizational, industrial, and cultural history in order to best prepare for crisis.

Theory development and model creation are needed to develop the anticipatory orientation with guidance for practitioners and researchers. Public relations, communication, rhetoric, business management, emergency response, and a host of other disciplines can come into play as scholars seek to organize and expand work on the anticipatory perspective. One of the prominent initial works in this vein was the creation of the anticipatory model (Olaniran & Williams, 2001). This work will be briefly reviewed below, as it serves to set the stage for the chapters that follow in this book.

The Anticipatory Model

The anticipatory model of crisis management was initially penned to provide assistance to public relations practitioners and researchers who must deal specifically with the threat of technological crises. The model necessitates an initial understanding and acceptance that technological crisis is an ever-present risk. Organizational paranoia of technological crises is justified, as previous research has demonstrated; it accounts for a significant percentage of reported crises (Mitroff, Pauchant, & Shrivastava, 1989). Adapters of technological innovation can easily fall prey to the perception that the materials and design are fail-safe. This lulling of an organization into a sense of ill-founded complacency harkens back to Kenneth Burke's (1961) warnings of a cultural movement toward Helhaven. Burke warned that as we continue to develop and depend on technology, we perceive ourselves moving closer to a heavenly state. However, the opposite may actually be more accurate. Incorporating the anticipatory orientation requires one to accept that greater dependence on technology leads to greater risk. Technology is never fail-safe. Even if it was, the human component of technology use is certainly not infallible.

Expectation and enactment are introduced to put the anticipatory model in motion. Expectations are the assumptions an individual, or a group, makes about an object, event, or decision. In the case of the model's original focus, the assumptions would be those made about technology. As Olaniran & Williams (2001) noted, "Assumptions by decision makers about technologies represent a critical starting point in crisis prevention" (p. 490).

These starting points might include assumptions that a new computer system cannot be compromised, proper installation assures proper functioning, or sufficient training for users will guarantee the prevention of operator error. A very general and unstated technological assumption may hold that the building housing the new technology is safe from natural disaster or terrorist attacks.

The anticipatory model is not limited to technological crises. For example, a financial services organization could be struck with crisis if it discovered advisors were misleading their clients. In this case the original expectation might have held that the hiring process, FBI background check, and months of training would prevent unscrupulous individuals from becoming advisors in the organization. Clearly, all organizations operate with a certain level of expectations present. While some expectations (e.g., training will prevent human error) may be easier to identify, others may be more difficult (e.g., our building is safe from a terrorist attack).

Enactment encompasses actions and consequences of actions put into motion. Olaniran and Williams (2001) noted, "Enactment constitutes a retrospective sense making process, such that the notion of anticipation (of crisis) in and of itself is an action given that it determines the choice that an organization makes based on derived information" (p. 490).

Enactment consideration follows expectations in the anticipatory model. Enactments are based on traditional crisis and risk assessments along with, or framed by, increased awareness of expectations. Enactment actions may include instituting preventative measures, corrective actions, and response strategies. The understanding and implementation of expectations alone is an enactment process.

It must be remembered that enactment includes not only the action, but also the results or consequences of that action as well. For example, the manufacturer of a weight-loss drink might factor in (anticipate) the likelihood that some consumers will experience ill effects from the product. Following this anticipation, the company can adjust labeling information and prepare response strategies for when complaints are offered. The anticipatory model would recognize, for example, the prepared response strategies as an action. However, the model would also have public relations personnel attend to the likely feedback from those responses. If the prepared response is to indicate the low number of side effects found in test results from the drink and to draw attention to the label warnings, those actions would probably not end the conversation with those registering complaints or the media and other stakeholders following up on the complaint. Instead, all interested parties would likely have some type of reaction or rebuttal to the prepared responses. These would be attended to when using the anticipatory model.

Expectation and enactment decisions must also be considered within the relevant internal and external foci. This represents continued movement within the model toward a more complete understanding of an organization's crisis vulnerabilities.

The internal focus draws public relations personnel to issues of rigidity and control. "Rigidity consists of the degree of inflexibility that is built into a particular action or process" (Olaniran & Williams, 2001, p. 492). Expectations (anticipations) and enactments made in crisis preparation will exist with a certain degree of rigidity, and that degree can change over time. An unstated expectation about the safety of a building may initially have low rigidity because few people think about that concern without a recent episode to suggest otherwise. People do not discuss the safety of the building, and regulations regarding safety are kept to a minimum. This situation would foster low rigidity in terms of how flexible safety procedures are followed and the concern among internal and external environments. However, the more safety procedures are discussed, or the more threats of possible terrorist acts are brought to public attention, the more rigid safety regulations will become. For example, throughout the United States there was a significant change in rigidity (flexibility) of safety procedures at airports and many public buildings from pre-9/11 to post-9/11. After 9/11 greater public and media attention was given to security procedures. This made the procedures more public and, therefore, more rigid.

The anticipatory model moves crisis practitioners to assess the level of rigidity that would exist in their crisis communication plans. Practitioners have to determine to what degree their actions (acts and statements) are codified by pre-existing knowledge or plans. If a local industry makes public their plans for how they would respond in case of a chemical spill, then they have high rigidity in their decision making and communication options. If that local industry has not made public how it would respond to a case of internal strife (e.g., strike, worker protest, or violence), then there is less rigidity in their actions and communication efforts.

Control refers to the degree to which individuals (usually crisis team members) can exercise decision making and implement actions in the face of a crisis. This consideration moves crisis team leaders and top management to determine whether crisis response will be more centralized or de-centralized. The decision here is a complicated one that will affect how well expectations and enactments are implemented at the outset of a crisis-triggering event.

There is reason to lean toward either greater or lesser control. Greater control is beneficial in that the importance of a unified message and an absence of misinformation are critical in crisis response, particularly in the initial response. If there is less control, decisions could be made and statements released that come back to haunt the organization. Initial reports of exorbitant death figures following a natural disaster provide one example of the dangers of less control. These occurrences make an already terrible situation even worse and heighten the concern of those who know people in the area of the disaster. It also reflects poorly on the work of the emergency response personnel, although they are often not the sources of the misinformation.

Less control is beneficial, as it can lead to the crisis response plan being enacted more quickly. One of the many issues with the crisis response during the *Exxon Valdez* tanker spill is that local personnel did not immediately approve the use of chemicals that would help break down or disperse the oil. According to procedure, they had to wait for permission from another authority to release the chemicals. This level of control slowed down the response process, worsened the spill, and damaged the image of all involved. Furthermore, crisis personnel who feel they have greater ability to react to crisis-triggering events (less control) tend to be more vigilant and feel a sense of empowerment (Weick, 1988). Olaniran and Williams (2001) address the dilemma over degree of control by noting, "It is recommended that a small span of control be used in the anticipatory crisis management model due to the interdependent and nonroutinized nature of tasks" (p. 493).

Finally, the anticipatory model must also take into consideration the relevant and ever-present external focus. Organizations do not exist in a vacuum. There will always be external factors affecting crisis anticipation and planning according to organization/industry type, location and culture, personnel, and organization size. Primary external considerations include legal issues, social/cultural considerations, economic factors, political issues and priorities, and media relations.

For example, consider the same crisis striking two different organizations in the same town: A business patron pulls out a gun, takes people hostage, and injures three people. One hostage situation takes place in an adult-oriented business with dancers and alcohol sales. The other is in a family-owned restaurant. One can assume that the external environment for these two organizations will be quite different. In some cases, the legality of adult-oriented businesses in specific locations come into question. It is also not uncommon for an element of protest against the business to exist in the local community. This protest may include political reaction against the business. These external factors will greatly alter the expectations and enactments the business owners can utilize in preparation for, and response to, such a crisis. Crisis responses of victimage, ingratiation, scapegoating, and denial (Benoit, 1995) may be plausible for the restaurant owner but will likely be unavailable to the owner of the adult-oriented business.

The anticipatory model of crisis management requires the crisis practitioner to think beyond the here and now. The model asks instead that crisis planning look inside and outside the organization to see what constraints are imposed by management style and the external environment that will alter crisis planning and crisis plan enactment. The model also affords a more challenging assessment of what expectations (assumptions) the organization holds about its crisis preparedness and vulnerabilities as well as what enactments will be utilized and the consequences of those actions. The following will help illustrate the need for the anticipatory model. We will look briefly at one industry that can benefit from the anticipatory orientation.

Coal Mine Crises

Mining accidents and disasters are a recurring event in United States and international crisis response. The unfortunate reality is the mining industry (coal and mineral) includes many dangerous work responsibilities. We continue to see mining crises despite continued improvements in safety technology and procedures. The inevitability of crisis is certainly apparent in the mining industry.

According to the United States Mine Rescue Association (Historical data, 2010), there have been 17 mine disasters since 1976. The number of mine disasters has decreased since the early 1900s. From 1951 to 1975 the total was 44 disasters; from 1926 to 1950, there were 170 disasters; and from 1901 to 1925, there were 356 disasters. However, when a new mining incident happens, public and media attention is rarely drawn to improvements in the industry and the decrease in the number of accidents. Instead, mine collapses are reported as the most recent in a continuing line of accidents.

The 2010 Chilean mine entrapment of 33 miners is a stark contrast to how mining incidents are usually portrayed in the public forum. Because of the dramatic rescue and continual stories of the miners' efforts to survive, the narrative focused largely on that mine and those miners. Media and public attention was drawn to and satisfied with reports on how the miners maintained life in a safe area of the mine. The world watched as we learned about the 33 individuals, their lives, and how they were able to communicate with those working to rescue them. The rescue efforts, while difficult, were explained in a simple manner. The plans were even conducive to computer-generated imaging to show how holes would be drilled and the miners extracted. The fact that all 33 trapped miners survived only helped enhance the story.

The Chilean mine crisis was significantly different than most mine collapses. The more typical portrayal of a mine disaster is that it is the latest in a line of disasters that plague the industry. Despite a significant reduction in the number of mining disasters, they strike frequently enough that the most recent incident serves as a reminder of previous incidents. For example, the 2007 Crandall Canyon mine accident in Utah stirred recent public memory of the Sago mine disaster of 2006 in West Virginia, which claimed the lives of 12 people. Generally, each new mine incident is reported in the context of all mine incidents. This serves to keep previous accidents alive in the public consciousness and to make the industry as a whole look unresponsive to safety issues.

Two other factors hinder the efforts of crisis management and communication personnel in responding to mine accidents. Incidents outside the United States are also frequently reported with great detail and time devoted to the story. These incidents on occasion have been of even greater magnitude than domestic mine crises. The 2009 XinXing coal mine disaster in China claimed 105 lives, and the 2010 Pike River coal mine explosion in New Zealand claimed another 29 lives. While they happened outside the United States, media coverage of these incidents reminds the public of similar mine incidents inside the United States. Further complicating the issue is that mines receive frequent inspections. These inspections do not typically result in an "all-clear" proclamation. Instead, violations are recorded, sometimes in great number, and made part of the mine's record. Inevitably, when a crisis hits a mine, there will be a paper trail of previous safety violations. Justified or not, the existence of these safety records will create an appearance of negligence or lack of concern for miner safety.

The mining industry would benefit greatly from the anticipatory perspective. Mine crises are predictable in that there is a long history to show what type of disaster would strike, if one does. The expectations about what could fail, what could go wrong, and to some degree what those consequences and risks would be are more limited and predefined than in most other industries. The research on mine safety and accidents is extensive, and the range of triggering events for crises can be focused on technological error and possible human error.

Pharmaceutical companies, for example, face crisis potential in drug testing, production, packaging, distribution, sales, and misuse by physicians or consumers. Crisis can happen at any point along the line in the production, sale, or use of each drug. The same can be said for companies in the toy industry. Efforts to anticipate possible crisis-triggering points can certainly be successful, but the range of possibilities exceeds that of the crisis industry. There are fewer anticipatory variables to account for in the mining industry.

Celebrity individuals and their images serve as another example of the differences between the coal industry and others in terms of expectations. Celebrity images can fall prey to a range of crisis triggers. The celebrities themselves can be the trigger for a crisis stemming from statements made (e.g., Don Imus), performance (e.g., Michael Richards), unflattering glimpses into their personal lives (e.g., Tiger Woods), or political beliefs. However, the crisis can emanate from other sources, such as work done by a company that the celebrities endorse (Nike and Coca-Cola for workplace allegations) or those who assist celebrities or serve as part of their entourage (Puff Daddy/P. Diddy Combs). Crises can also stem from members/groups of the public who dislike celebrities' work or disapprove of their actions, other celebrities who bring their names into public light in a negative way, and cultural groups who disapprove of the way their culture is represented by the celebrity (e.g., Angelina Jolie and Bosnian war victims).

In comparison, crisis concern for the coal industry is largely relegated to the extraction of the product. There is much less concern with crisis potential after the coal has been extracted from the mines. The misuse of coal by the consumer, for example, has less crisis potential for coal industry management than misuse of drugs by the consumer has for pharmaceutical industry management.

Given the predictability of coal mining crises, crisis management personnel can be more focused in the assumptions made during crisis planning. They know with much greater specificity what the crisis-triggering events are likely to be, the type of crisis that is most likely to occur, and the potential for real damages that will ensue. Mine industry crisis personnel can also make more specific assumptions about how disasters will be reported in the media and how the general public will initially respond to news of the crisis. This is not to say that identifying and generating expectations for coal industry crisis preparations would be an easy task. However, it does suggest that the anticipatory model would seem very conducive to the work needed in crisis planning for the coal industry.

The same history and predictability of coal mine crises would assist crisis practitioners using the enactment mechanism of the anticipatory model. Mining disaster history has clear examples of how the public responds to crisis and different strategies employed by organizations responding to the crisis. Coal mining crisis personnel should be able to determine—again, to a greater extent than in many other industries—what actions they can prepare for and even what communication strategies they would employ. For example, previous mine disasters show that the presence and continued visibility of the owner and/or mine operation president at the site of the crisis, as well as frequent updates delivered by this individual, are received more positively than when that high-ranking figure is not present.

Enactments in coal mining crisis preparation would include assumptions made about crisis vulnerability and risk, actions taken to keep stakeholders and the public informed of rescue and recovery efforts, and communication response strategies. The consequences of these enactment actions will be more predictable because of the limited range of assumptions about risks and triggering events.

Conclusion

The coal industry provides just one example of how the anticipatory orientation and model can assist practitioners with development of crisis management planning. The anticipatory model is one option for incorporating a perspective that recognizes the need to further the progression from a response-oriented approach to crisis management to one that devotes attention to anticipation of the inevitable.

References

Barton, L. (1993). *Crisis in organiz*ations. Cincinnati, OH: College Divisions South-Western.

———. (2001). *Crisis in organizations II.* (2nd ed.). Cincinnati, OH: College Divisions South-Western.

———. (2007). *Crisis leadership now.* Boston: McGraw-Hill.

Benoit, W. L. (1995). *Accounts, excuses, and apologies: A theory of image restoration.* Albany: State University of New York Press.

Benson, J. A. (1988). Crisis revisited: An analysis of strategies used by Tylenol in the second tampering episode. *Central States Speech Journal*, 39, 49–66.

Burke, K. (1961). *Attitudes toward history.* Boston: Beacon.

Coombs, T. (1995). Choosing the right words: The development of guidelines for the selection of the "appropriate" crisis-response strategies. *Management Communication Quarterly*, *8*, 447–476.

———. (2007). *Ongoing crisis communication.* (2nd ed.). Los Angeles: Sage.

Fink, O. (1986). *Crisis management: Planning for the inevitable.* New York: AMACOM.

Heath, R. L., & Millar, D. P. (2004). A rhetorical approach to crisis communication: Management, communication processes, and strategic responses. In Millar, D. P. & Heath, R. L. (Eds.). *Responding to crisis: A rhetorical approach to crisis communication* (30–39). Mahwah, NJ: Lawrence Erlbaum.

Historical data on U.S. mine disasters. (2010). www.usmra.com

Lerbinger, O. (1997*). The crisis manager: Facing risk and responsibility.* Mahwah, NJ. Lawrence Erlbaum.

Mitroff, I. I. (1986) Prevention: Teaching corporate America to think about crisis prevention. *Journal of Business Studies*, *6*, 40–48.

———. (2003). *Crisis leadership: Planning for the unthinkable.* San Francisco: Wiley.

Mitroff, I. I., Pauchant, T., & Shrivastava, P. (1989). Can your company handle a crisis? *Business & Health, 7*(5), 41–44.

Olaniran, B. A., & Williams, D. E. (2001). Anticipatory model of crisis management: A vigilant response to technological crises. In R. L. Heath (Ed.), *Handbook of Public Relations*, (pp. 487–500). Thousand Oaks, CA: Sage.

Pauchant, T. C., & Mitroff, I. I. (1992). *Transforming the crisis-prone organization: Preventing individual, organizational, and environmental tragedies.* San Francisco: Jossey-Bass.

Perrow, C. (1984). *Normal accidents: Living with high risk technologies.* New York: Basic Books.

Seeger, M. W., Sellnow, T. L., & R. R. Ulmer (2001). Public relations and crisis communication: Organizing and chaos. In R. L. Heath (Ed.), *Handbook of Public Relations* (pp. 155–165). Thousand Oaks, CA: Sage.

Weick, K. (1988). Enacted sensemaking in critical situations. *Journal of Management Studies*, *25*, 305–317.

Williams, D. E., & Olaniran, B. A. (1998). Expanding the crisis planning function: Introducing elements of risk communication in crisis communication practice. *Public Relations Review*, *24*, 387–400.

Chapter 2

Issue Management as a Strategic Aspect of Crisis Prevention

Tony Jaques

The author takes the long-standing crisis management concern for issue management and identifies how it can and should play a role in pre-crisis planning. Within this chapter, audits, preventive maintenance, emergency response, environmental scanning, risk assessment, and issue prioritization are all addressed in terms of effective crisis prevention and planning. Material is presented in this chapter in a manner that is readily intelligible to the scholar and the practitioner charged with crisis preparation and prevention responsibilities.

Introduction

The concept of pre-crisis management—when proactive steps can and should be taken to help prevent a crisis before it happens—is not new. Yet after more than two decades, the pre-crisis phase of crisis management has still to be fully characterized and defined. That in turn inhibits the emergence of an optimal management discipline to help organizations navigate this critical area.

Issue management holds the potential to produce such an effective strategic framework for crisis prevention, but it is held back by lack of substantive consensus at the academic level and by a lack of agreed approaches and practical tools for practitioners.

Although the general idea of crisis management has been present for generations, much of its early focus was on natural disasters, leading to societal crises, and "international crises"—which were often taken to be politico-military events. The development of crisis management as a formal business/organizational activity is considerably more recent.

The much-mythologized Tylenol crisis of 1982 (Jaques, 2009c) is often cited as the single most important event in the birth of the discipline in the United States. Mitroff (2001) concluded: "The modern field of crisis management is generally acknowledged to have started with the Tylenol poisonings" (p. 3), and this view is echoed by other authorities, including Heath and Palenchar (2009). On the other side of the Atlantic, Falkheimer and Heide (2006) argue that the singular event that similarly galvanized the development of crisis management in Europe was the Chernobyl disaster of 1986.

This timing is important because it closely parallels the introduction of issue management, coined by Howard Chase in the inaugural issue of his publication *Corporate Public Issues and Their Management* (1976). However, issue management as a formal business discipline did not gather real momentum until the early 1980s, boosted by the publication of Chase's seminal book, *Issue Management: Origins of the Future*, in 1984. Significantly, one of the groundbreaking books on crisis management as a business discipline appeared just two years later: Steven Fink's *Crisis Management: Planning for the Inevitable* in 1986.

The link between these parallel developments has had a very important impact on growth of the process approach to crisis management and also on the positioning of issue management as a key tool for the pre-crisis phase of the process.

Issue Management as a Crisis Prevention Discipline

Issue management as originally conceived was very much a business-based discipline designed to enable the corporation to participate in and not simply respond to public policy (Chase, 1976). This strong focus on public policy and politics is still very relevant in some areas of issue management practice (as defined, for example, by the Public Affairs Council in Washington, D.C.) and has retained currency among some academics. For example, Heath and Coombs say: "Issue management is the management of organizational and community resources through the public policy process to advance organizational interests and rights by striking a mutual balance with those of stakeholders and stakeseekers" (2006, p. 262).

However, even from the earliest days, issue management was also perceived as an early warning or pre-crisis mechanism, closely aligned with strategic planning. In this broader, proactive context, issue management can be defined as a formal discipline, which "attempts to minimize surprises which accompany social and political change by serving as an early warning system for potential environmental threats, and attempts to promote more systematic and effective responses to particular issues by serving as a co-ordinating and integrating force within the corporation" (Wartick & Rude, 1986, p. 124).

Similarly, Renfro (1993) argued: "The over-riding goal of an issue management function is to enhance the current and long term performance and standing of the corporation by anticipating change, promoting opportunities and avoiding or mitigating threat" (p. 107).

This strongly corporate, pre-emptive approach was championed very early by the business leader Archie Boe, who was closely aligned with the founders of the discipline. He described issue management as a response to what he called an unprecedented battering of business:

> Crisis and post-crisis management have been the only responses open to chief executive officers until recently. During the past few years, however, intensive study of the processes and forces which have brought about these changes has resulted in the belief that business can move to a pre-crisis management posture and participate in the public policy process that resolves these larger demands on business. The pre-crisis management approach is called issues management and is an important management tool available to today's business leaders. (Boe, 1979, p. 4).

Yet the potential role of issue management specifically as a pre-crisis or crisis prevention tool did not mature until after the emergence of organizational crisis management itself as a recognized independent discipline in the mid-1980s.

Among the first academics to formally develop the broad concept of issue management as an integral element of crisis prevention were González-Herrero and Pratt (1995), who proposed what they called "a four phase conceptual model for the effective management of a crisis—before or whenever it occurs" (p. 26), which they said was guided by three principles: issues management, planning-prevention, and implementation. They proposed that the first main phase of the crisis management process is issue management, during which stage "the company's resources are aimed at identifying any threatening issue and influencing its course" (p. 28).

González-Herrero and Pratt (1996) subsequently characterized this four-step model as "a proactive, symmetrical crisis-management process" and argued:

> The first task a crisis-communication manager must undertake is the one that falls under issue management. The activities that should be accomplished at this point represent the highest level of proactivity that an organization can offer. Issues management theory enables the practitioner and the organization to lead the environment in search for troublesome issues—before they are obligated to look for them. (p. 89)

While this approach was built on the premise that crises progress through a recognized life cycle, it did not fully evolve until the further development of crisis management itself. The traditional event approach presents crisis management as a largely tactical activity focused on incident response—what to do when a crisis occurs and how to prepare for it in case it happens. By contrast, the process approach presents crisis management as a contributory function in a comprehensive sequence of management activity extending well before the event. One of the early American champions of the process approach was Shrivastava, who wrote: "Crises are not events, but processes extended in time and place" (1995, p. 2). This concept was later taken up strongly in Europe, where leading academics further developed the concept and explored in detail the two approaches and their relative merits (for example, 't Hart, Heyse, & Boin, 2001; Forgues & Roux-Dufort, 1998; Roux-Dufort, 2007).

Typical of this movement is Roux-Dufort's characterization of a crisis as "an accumulation of organizational imperfections." Describing the new paradigm, he concluded: "The process approach to crises…leads us to see the triggering event as the factor that reveals a preexisting dynamic of crisis. In other words, what the event approach sees as the crisis (i.e., the triggering event) the process approach only sees as the amplifier of a process that started long before. Therefore the crisis grows more intense and visible" (2007, p. 228).

Advancement of this process approach helped turn attention to the other process-driven disciplines, which form part of the pre-crisis phase and, in particular, to the integral relationship between issue management and crisis management. Specifically, the process approach to crisis management gave fresh impetus to recognition of issue management as a strategic aspect of crisis prevention, emphasizing that not only do crises follow a distinct path before the triggering event but also that effective managers can and should take pre-emptive action early in the cycle.

Developing the Concept of a Cycle

The concept of applying a phased or cyclical construct was an early feature of the development of both issue management and crisis management.

At the beginning of issue management, Howard Chase and his colleague Barrie Jones published the first formal issue management process diagram (Chase & Jones, 1977), which was comprised of five steps:

1. Issue identification, leading through theory and research to
2. Issue analysis, followed by judgment and priority setting to reach
3. Issue change strategy options, which in turn lead through policy for change strategy to
4. Issue action program, then leading on to
5. Accomplishment of goal/evaluation of results, and looping back to the first stage.

Coombs and Holladay (2007) argue that this is still regarded as "the most influential issue management model" (p. 81). It laid the foundation for what became known as the issue life cycle, which is a graphical representation of the fact that issues tend to progress and deteriorate over time, becoming harder and more costly to manage in the face of increasing public and legislative attention.

An early example of the issue life cycle was the work of Meng (1992), whose model showed that issues moved through five phases of a life cycle: potential, emerging, current, crisis, and dormant. Importantly, Meng's model unambiguously presented issue management as part of a continuum up to crisis and beyond. This concept has been reinforced in many other life cycle models, illustrating issue management in the pre-crisis or crisis prevention phase, with a crisis sometimes characterized as a failure of issue management or an issue which wasn't properly managed.

In a similar way to the issue life cycle, many crisis management models also propose a phased structure. The crisis management pioneer Barry Turner (1976) introduced the idea of predictability and developed the concept of pre-crisis "incubation," which was later absorbed into organizational crisis management. For example Fink's 1986 crisis cycle was a four-stage model: prodrome, acute crisis, chronic crisis, and resolution. Fink himself said the prodromal stage was occasionally referred to as the pre-crisis stage, but he added: "…usually that appellation is used after the acute crisis has hit, when, in retrospect, people look back at a series of events and point to something as a pre-crisis" (p. 21).

Many other sequential crisis models have subsequently been developed by a number of researchers including Elliott, Smith and McGuinness (2000), Harrald, Marcus, and Wallace (1990), Kovoor-Misra (1995), Pearson and Mitroff (1993), and Smith (1990). But unlike Fink, who suggested that recognition of a pre-crisis usually reflects the wisdom of hindsight, these and other later scholars say the early phases should constitute genuine, planned steps to prevent crisis and to reduce the risk of an issue developing into a crisis. Or, as Smith described it: "To be effective, crisis management should, almost by definition, include systematic attempts to prevent crises from occurring" (2005, p. 312).

In other words, the modern conceptualization of crisis management is much more than just crisis response. Penrose (2000) used the phrase "proactive crisis management" and took this approach further, arguing that pre-crisis and post-crisis activities cluster together and should be considered in aggregate rather than as separate sets of activities.

Applying Issue Management to Crisis Prevention

One of the first models to build on Penrose and implement this fully integrated approach is Jaques (2007a), who presented a non-linear cyclical process with issue management as part of both the pre-crisis and post-crisis clusters of activities. While issue management as part of the post-crisis cluster is explored elsewhere (Jaques, 2009b), the pre-crisis phase is divided into two distinctly defined constructions, namely crisis preparedness—the largely tactical and mechanistic activities required to get ready for a crisis—and crisis prevention, which includes both tactical and strategic activities designed to help reduce the risk of a crisis occurring in the first place.

The tactical activities of crisis prevention shown in this model include routine audits, preventive maintenance, and effective emergency response. In respect to this last item, not every crisis is triggered by an emergency, but emergencies badly handled can and do lead to crises. Emergency response is often regarded as a stand-alone tactical procedure, but it can in fact be an important element of crisis prevention.

Also included within this model's crisis prevention phase are more strategic activities, such as environmental scanning, risk assessment, and issue prioritization, which contribute to issue management as a strategic element of crisis prevention. Traditional strategic planning is another factor in this phase, and issue management and strategic planning have much in common. But it is essential to recognize that the two disciplines also have significant differences, and those differences should not be allowed to blur for the sake of convenience. (For a detailed analysis of the relationship between issue management and strategic planning, see Jaques, 2009a.)

Sadly, practice suggests that, despite broad acceptance that crisis management extends well before any triggering event, much of the focus remains in reactive mode. Forgues and Roux-Dufort (1998) conceded that "while most of us appear to agree on the fact that crises are processes, we nevertheless often treat them as events" (p. 3).

Similarly, at the applied level, Pauchant and Mitroff argued that 90% of the crisis management literature focuses primarily on what to do in response to a triggering event and how to pick up the pieces afterward. They described this reactive mode as crash management:

> Crisis management is not the same as crash management—what to do when everything falls apart. Obviously this is important, but it is only one part of total crisis management effort. Here we focus not only on crash management—what to do in the heat of a crisis—but also on why crises happen in the first place and what can be done to prevent them. (Pauchant & Mitroff, 1992, p. 11)

When crisis management is seen in the reactive crash management mode, it is inevitable that any concept of crisis prevention will focus primarily on control measures designed solely to minimize damage. But if crisis prevention is seen as part of a much broader concept of pre-crisis management, the established and well-defined discipline of issue management has a great deal to offer, utilizing many of the same tools and techniques.

Applied during the pre-event stage, there are four broad areas where issue management can contribute to crisis prevention:

1. Proactively addressing underlying systemic causes of potential crisis
2. Establishing effective signal detection mechanisms
3. Properly identifying stakeholders and their perspectives
4. Learning and unlearning on an ongoing basis

Proactively Addressing Underlying Systemic Causes of Potential Crisis

Systemic failure can exist in a number of forms. One form concerns signal detection, intended to both recognize and respond to warning signs or specific events or trends (discussed below). Systemic management failure might relate to failure to put in place processes and structures designed to facilitate such recognition and response. Another aspect of the same risky deficiency might be an organization accepting or even promoting management systems and structures that limit signal detection, inhibit upward reporting, and discourage dissent.

High-profile examples of such systemic failure would include the early warnings of the Ford Explorer/Firestone tire crisis, which emerged outside the United States and were "explained away" (Bridges, 2004); the Mitsubishi Motor quality scandal, where evidence of recall faults was deliberately suppressed (Hagiwara, 2007); the *Challenger* disaster, where NASA management systems worked to block dissenting views (Esser & Lindoerfer, 1989; Starbuck & Milliken, 2007); and the Société Générale loss of $7 billion (4.9 billion euros) from the activities of a rogue trader, where an independent panel found that the French bank failed to act on 75 red flags or early warnings over a period of 18 months (Clark, 2008).

It has been said that while systemic problems are easy enough to identify, solutions are much harder to find (Mitroff & Pauchant, 1990). Although much has been written about the inevitable existence of red flags and how easy it is to miss them, there has been only limited scholarship on the formal processes to ensure this does not occur. The issue management approach provides an effective tool to address such underlying problems because it requires going beyond the obvious systemic risks, which are intrinsic to particular industries, such as contamination scares for food producers, fires and spills at an oil refinery, or security and fraud risks at a financial institution.

Moreover, underlying systemic potential issues may apply across a wide variety of business sectors—beyond industry-specific risks and beyond generic problems such as floods, fires, power outages, infrastructure breakdowns, and natural disasters. These broader systemic causes might typically include uncompetitive behavior, sexual or racial harassment or discrimination, executive dishonesty, takeovers and mergers, layoffs, new technology, off-shoring, and management misconduct. They are often areas where management systems and structures constitute a crucial vulnerability, particularly where there is an implicit or explicit desire to avoid sensitive issues that might prove embarrassing to management.

In early research, Dunbar and Goldberg (1978) studied 20 corporations enmeshed in crises and concluded that the chief executives generally surrounded themselves with *yes-sayers* who voiced no criticism and who deliberately filtered out warnings from middle managers who correctly saw that their corporations were out of touch with the market realities. Stocker (1997) reinforced this view: "Keep in mind that top management, by definition, is the least informed group in the company when it comes to bad news. Nothing moves more slowly than bad news running up a hill, a very steep hill" (p. 192).

More specifically, following the disastrous explosion and refinery fire at BP Texas City in August 2005, a researcher found a culture at BP in which many top people knew of problems, but few would speak up. "Only good news flowed upward," he reported. "No one dared say the wrong thing or challenge the boss" (cited in Hopkins, 2009, p. 108).

However, Brown (2002) suggests that the greater systemic challenges for management structures are not lack of good information but (1) lack of imagination; (2) faulty or inadequate analysis; (3) failure to see the big picture; and (4) failure to link information to action.

This deficiency was well captured by the final report of the United States House of Representatives Select Bipartisan Committee (2006) that was set up to investigate the preparation for and response to Hurricane Katrina. The committee's 520-page report was aptly titled *A Failure of Initiative*, and the committee concluded: "While there was no failure to predict the inevitability and consequences of a monster hurricane—Katrina in this case—there was a failure of initiative to get beyond design and organizational compromises to improve the level of protection afforded" (p. 97).

Issue management, with its deliberate action orientation and no-fault objective analysis, can help transform such information and intent into effective action.

Establishment of Effective Signal Detection Mechanisms

The link between information and action lies at the heart of effective signal detection mechanisms, and there is an obvious close relationship between systemic causes of failure to recognize potential crises and failure to detect signals. Many authors have identified that most, if not all, crises are preceded by clear warning signals and that these are frequently ignored. For example, James and Wooten (2005) concluded "smoldering crises nearly always leave a trail of red flags or warning signals that something is wrong. Unfortunately, these signals often go unheeded by management" (p. 143). The crisis scholar Ian Mitroff went even further to assert that in every crisis he had ever studied, there were always a few key people on the inside of an organization or on its edge who saw the early warning signs and tried to warn their superiors. "In every case the signals were either ignored or blocked from getting to the top or having any effect" (Mitroff, 2002, p. 20).

Supporting this observation is the work of the Institute for Crisis Management (www.crisisexperts.com), which has been tracking business crises in the news media since 1994. Their data consistently show that two thirds of crises are not sudden events but are categorized as smoldering crises—events which should have and could have prompted early intervention.

A typical case is the notorious McDonald's hot coffee incident of 1992 when an elderly New Mexico woman was scalded by hot coffee spilled in her lap. Analyzing this case, González-Herrero and Pratt (1996) reported that the company had received at least 700 complaints of coffee burns in the previous decade. "McDonald's knew that its coffee was among the hottest—if not the hottest—in the industry. It seemed the fast-food chain also knew its coffee sometimes caused serious burns; however it did not consult experts about the issues. Both these issues were seemingly overlooked" (p. 87).

While the focus here is primarily on organizational crises arising from failure to detect signals, a telling example is provided by consultant Regina Phelps (2005) in an analysis of potential crises facing the United States. Phelps reported that in 2001 the Federal Emergency Management Agency (FEMA) formally identified the three most probable catastrophes facing the country: a terrorist attack in New York City, a catastrophic earthquake in San Francisco, and a hurricane and levee break in New Orleans. The first of these happened just four months later on September 11, 2001, and Hurricane Katrina destroyed New Orleans in August 2005.

In July 2004, a massive simulation was staged by United States federal and state emergency planners in which the fictional Hurricane Pam struck New Orleans in a catastrophic scenario almost identical to what occurred in reality only 12 months later. Yet Homeland Security Secretary Michael Chertoff later argued that the real Hurricane Katrina was "particularly unpredictable" and "exceeded the foresight of planners, and maybe anybody's foresight" (Lipton & Shane, 2005)

Whether it is natural disasters or corporate crises, it is clear that detecting signals and also responding to signals is no simple task. Mitroff, Harrington, and Gai warned that, despite the early warnings in advance of the actual crisis itself, the detection of such signals "remains an enigma for most organizations" (1996, p. 44).

Environmental scanning and signal detection is a core management process across many information systems, planning procedures, and decision-making techniques, such as strategic forecasting, contingency planning, issues analysis, and scenario analysis (Kash & Darling, 1998). Yet, while the concept of environmental scanning has broad applications, issue management took this further and developed sophisticated tools and techniques for objective methods to identify, categorize, and prioritize issues and potential crises and the informational context in which they lie (for example, Dyer, 1996; Lauzen, 1995). Although the specific methodologies may vary, the consistent elements represented here are objectivity, independence, and the link to action.

A case which has been described as an archetypal example of failure to link information to action is the 1995 disaster that befell Baring's Bank, a British financial institution in business since 1793 (Swartz, Elliott, & Herbane, 1995). While the bank was eventually destroyed by the activities of a single rogue trader, it was evident very early that the bank's internal control systems failed to catch signals such as increased trading activity, the extensive use of leverage, and escalating volume of trades.

More recently, in the wake of the $65 billion Bernie Madoff investment scandal, U.S. Security and Exchange Commission Inspector General David Kotz reported that the agency missed "numerous red flags" from 1992 until the fraudster was arrested in December 2008. Kotz said five separate failed investigations into the affair had been bungled, and he called for major organizational change (Stout, 2009).

Properly Identifying Stakeholders and Their Perspectives

While stakeholder identification and analysis are core public affairs activities, the processes developed within issue management have enhanced this skill, particularly in relation to the role and response of opponents, including non-government organizations (NGOs) and social and political activists. Heath is a leading proponent of issue management as a discipline "to reduce friction and increase harmony between organizations and their publics" (for example, Heath, 2005, p. 460), and the importance of such understanding as an element of crisis prevention is illustrated by a number of high profile classic cases where inadequate understanding of stakeholder potential led to serious failure.

One of these was Monsanto's dramatic misjudgment of the state of European public opinion over biotechnology, by which the company "wrecked an entire industry, as well as its own brand" (Larkin, 2003, p. viii). Speaking in the wake of the disaster, Monsanto CEO Bob Shapiro said:

> We started with the conviction that biotechnology was useful and valuable but we have tended to see it as our task to convince people that we were right and that people with different points of view were wrong. We have irritated and antagonised

> more people than we have persuaded. Our confidence in biotechnology has been widely seen as arrogance and condescension because we thought it was our job to persuade. But too often we forgot to listen. (Vidal, 1999)

Another company that famously misjudged the role of external parties as key stakeholders was Shell UK with its failed attempt to sink the disused oil storage buoy Brent Spar in the North Atlantic (Bakir, 2006; Entine, 2002). Not only did the company completely fail to understand the role of Greenpeace activists, who occupied the abandoned platform, but it also misread public and political opinion in Europe and the opposition of Shell's own European subsidiaries, which eventually forced abandonment of the plan. Shell UK Chairman and CEO Christopher Fay frankly admitted that although they had covered all the scientific and technical and legal angles, they "hadn't taken into account hearts and emotions, where people are coming from" (BBC, 1995).

The following year, Shell Chairman Cor Herkstroter conceded that the company failed to recognize that the issue was symbolic as well as technical.

> For most engineering problems there is a correct answer. For most social and political dilemmas there is a range of possible answers, almost all comprises. So starting off with a strong, scientifically grounded mindset, we tended to misjudge some of the softer issues and consequently made mistakes. We misread some of the situations. (Herkstroter, 1996, p. 103)

Such disastrous failure to properly understand stakeholder perspectives can also apply to much more traditional stakeholders, such as customers, as was shown in the notorious Intel Pentium chip case, where it was computer users whose voices were ignored. After weeks of denial and mounting criticism, Intel finally agreed that the computer chip fault was more serious than they had admitted and announced a recall, accepting a one-off charge against earnings of $475 million. Intel CEO Andrew Grove later conceded: "We got caught between our mindset, which is a fact-based, analysis-based engineer's mindset, and customers' mindset, which is not so much emotional but accustomed to making their own choice" (Carlton & Yoder, 1994, p. B1). In his analysis of this famous case, Hearit (1999) concluded: "It appears that in the case of Intel, an apology does not necessarily mean admitting guilt, but it does mean acknowledging that the customer is always right" (p. 300).

How commercial organizations can underestimate stakeholders and misunderstand their perspective is also well illustrated by a less well-known case involving the food giant Nabisco, which announced in 1996 that it was discontinuing 300 different brands including Crown Pilot, a low-selling large cracker distributed in only three coastal states of New England.

Local protesters in a fishing community in Maine challenged the decision, arguing that the crackers were "an integral element of the culture of an entire region" and that Nabisco was "a big corporation stripping a small community of its heritage" (Esrock, Hart, D'Silva, & Werking, 2002, p. 215). Following a cleverly orchestrated local campaign, the "David and Goliath" story went national, and Nabisco capitulated, reinstating the Crown Pilot Crackers amid what the company called "full contrition mode" (Esrock et al., p. 221).

Learning and Unlearning on an Ongoing Basis

Learning from crises is a subject in itself with well-established literature regarding the barriers to management learning after a crisis (Elliott, Smith, & McGuinness, 2000; Gibson, 2000; Roux-Dufort, 2000); how senior managers often don't learn from the crises of others (Jaques, 2008; Mitroff & Pauchant, 1990; Stern, 1997); and recognition that formal management learning processes such as crisis simulations are often limited to those areas where they feel comfortable (Kovoor-Misra, Zammuto, & Mitroff, 2000).

Twenty years ago, Shrivastava (1988) noted that the one thing more tragic than the crises which occur is the failure of organizations and organizational scholars to learn from them. This was later echoed by Stocker (1997) who commented that when you look at the majority of crises that occur, what happened should have been on or near the top of the list of possible events. "Why?" he asked. "Wasn't anyone prepared?" (p. 192)

In essence, ongoing crisis learning falls largely into two general areas: honest evaluation of what has happened in the past and open scanning, which takes information to the top. While "no-fault" learning from crises and near misses—through objective investigation and evaluation—is a self-evident way forward, there is no doubt that the upward flow of information is a major obstacle. Mitroff (2004) has suggested that one of the most significant barriers to effective crisis management is denial, while Nystrom and Starbuck (1984) argue that top managers should listen to and learn from "dissenters, doubters and bearers of warnings" to remind themselves that their own beliefs and perceptions may well be wrong. "Indeed, top managers should worry if they hear no such messages: long silences signal distortion, not consensus" (p. 60).

In this regard, Hopkins (2009) has discussed in detail what he calls mindful leadership:

> Mindful leaders do not rely on assurances from subordinates that all is as it should be. They do not assume that the system is functioning as it should be in theory. They know, or at least they fear, that there are problems lying in wait to pounce, and they use every means available to probe for these problems and expose them before they can impact detrimentally on the organization….Mindful leaders welcome bad news. They recognize that it is often difficult to convey bad news upwards and they develop systems to reward the bearers of bad news. (p. 120)

The challenge of effective scanning and upward communication was highlighted by Intel CEO Andy Grove in the wake of his company's faulty chip crisis:

> Most CEOs are in the center of a fortified palace, and news from the outside has to percolate through layers of people from the periphery where the action is. I was one of the last to understand the implications of the Pentium crisis. It took a barrage of relentless criticism to make me realize that something had changed and that we needed to adapt to the new environment. (Grove, 1996, p. 62)

Conclusion

Given a focus on crisis preparedness and crisis prevention, the systemic answers to each of the principal areas of management learning are not hard to see but, rather, much harder to implement. An effective preventive process basically requires capacity to scan the environment, gather information, assess and evaluate that information, and turn it into action. Issue management provides a framework for such an integrated process, which can operate across functions and into the highest levels of management.

It has been argued (Smith, 1990) that one of the reasons crisis prevention has been severely neglected compared to crisis management is because it raises fundamental questions about the nature of managerial style and organizational culture. Smith proposes that this is largely because practicing managers feel threatened by the changes such questions bring with them.

However, another reason is undoubtedly the lack of agreed, neutral processes to address such change. Issue management properly implemented is an inherently neutral process that permits risk and change to be presented and assessed objectively. This is particularly important in light of the fact that organizational issues are, by definition, problems where there is no black and white answer; where emotions rather than data often prevail; where the risks of failure are greatest; and which, if left unmanaged, have the potential to become crises and threaten the entire organization (Jaques, 2007b).

Issue management has always contained within its processes and its literature the essential elements to help prevent crises before they happen. But the emergence of the process approach to crisis management—and the ongoing evolution of issue management as an effective tool within the broader concept of crisis management—is a convergence of two developments that jointly offer a unique opportunity to formalize and systematize crisis prevention.

References

Bakir, V. (2006). Policy agenda setting and risk communication: Greenpeace, Shell and issues of trust. *Harvard International Journal of Press/Politics*, *11*(3), 67–88.

BBC (1995). The battle for Brent Spar, *BBC Public Eye*. Broadcast BBC2, 3 September, 1995.

Boe, A. R. (1979). Fitting the corporation to the future. *Public Relations Quarterly*, 24(4), 4–6.

Bridges, J. (2004). Corporate issues campaigns: Six theoretical approaches. *Communication Theory*, *14*(1), 51–77.

Brown, A. (2002). Avoiding unwelcome surprises. *The Futurist*, *36*(5), 21–23.

Carlton, J., & Yoder, S. K. (1994, December 21). Humble pie: Intel to replace its Pentium chips. *The Wall Street Journal*, pp. B1.

Chase, W. H. (1976). Objectives of CPI. *Corporate Public Issues and Their Management*, *1*(1),1.

———. (1984). *Issue management: Origins of the future*. Stamford, CT: Issue Action.

Chase, W. H.., & Jones, B. L. (1977). CPI presents. *Corporate Public Issues and Their Management*, *2*(14), 1–4.

Clark, N. (2008, February 21). Société Générale posts record loss on trading scandal, subprime exposure. *The International Herald Tribune*. Retrieved from http://www.iht.com/articles/2008/02/21/business/socgen.php

Coombs, W. T., & Holladay, S. J. (2007). *It's not just PR: Public relations in society*. Malden, MA: Blackwell.

Dunbar, R. L. M., & Goldberg, W. H. (1978). Crisis development and strategic response in European organizations. *Journal of Business Administration*, *9*(2), 139–149.

Dyer, S. C. (1996). Descriptive modeling for public relations environmental scanning: A practitioner's perspective. *Journal of Public Relations Research*, *8*(3), 137–150.

Elliott, D., Smith, D., & McGuinness, M. (2000). Exploring the failure to learn: Crises and barriers to learning. *Review of Business*, *21*(3),17–24.

Entine, J. (2002). Shell, Greenpeace and Brent Spar: The politics of dialogue. In C. Megane & S. J. Robinson (Eds.), *Case Histories in Business Ethics* (pp. 59–95). London: Routledge.

Esrock, S. L., Hart, J. L., D'Silva, M. U., & Werking, K. J. (2002) The saga of the Crown Pilot: Framing, reframing and reconsideration. *Public Relations Review*, *28*(3), 209–227.

Esser, J. K., & Lindoerfer, J. S. (1989). Groupthink and the space shuttle Challenger accident: Toward a quantitative case analysis. *Journal of Behavioral Decision Making*, *2*(3), 167–177.

Falkheimer, J., & Heide, M. (2006). Multicultural crisis communication: Towards a social constructionist perspective. *Journal of Contingencies and Crisis Management*, *14*(4), 180–189.

Fink, S. (1986). *Crisis management: Planning for the inevitable*. New York: American Management Association.

Forgues, B., & Roux-Dufort, C. (1998, May 26–27). Crises: Events or processes? *Hazards and Sustainability*. Conference conducted at Durham, UK.

Gibson, D. C. (2000). Firestone's failed recalls, 1978 and 2000: A public relations explanation. *Public Relations Quarterly*, *45*(4), 10–13.

González-Herrero, A., & Pratt, C. B. (1995). How to manage a crisis before—or whenever it hits. *Public Relations Quarterly*, *40*(1), 26–29

———. (1996). An integrated symmetrical model for crisis-communications management. *Journal of Public Relations Research*, *8*(2), 79–105

Grove, A. S. (1996, September 2). How we miscalculated: A minor flaw in its biggest chip awakened Intel to a new business reality. *Newsweek*, 60–62.

Hagiwara, T. (2007) The eight characteristics of Japanese crisis-prone organizations. In C. M. Pearson, C. Roux-Dufort & J. A. Clair (Eds.), *International handbook of organizational crisis management* (pp. 253–270). Thousand Oaks, CA: Sage.

Harrald, J. R., Marcus, H. S., & Wallace, W. A. (1990). The Exxon Valdez: An assessment of crisis prevention and management systems. *Interfaces*, *20*(5), 14–30.

Hearit, K. M. (1999). Newsgroups, activist publics and corporate apologia: The case of Intel and its Pentium chip. *Public Relations Review*, *25*(3), 291–308.

Heath, R. L. (2005). Issues management. In R. L. Heath (Ed.), *The encyclopedia of public relations* (pp. 460–463). Thousand Oaks, CA: Sage.

Heath, R. L., & Coombs, W. T. (2006). *Today's public relations: An introduction.* Thousand Oaks, CA: Sage.

Heath, R. L., & Palenchar, M. J. (2009). *Strategic issues management: Organizations and public policy challenges*, (2nd ed). Thousand Oaks, CA: Sage.

Herkstroter, C. (1996). Dealing with contradictory expectations: Dilemmas facing multinationals. *Vital Speeches of the Day*, *63*(4), 100–104.

Hopkins, A. (2009), *Failure to learn: The BP Texas City refinery disaster*. Sydney, Australia: CCH Australia.

James, E. H., & Wooten, L. P. (2005). Leadership as (un)usual: How to display competence in times of crisis. *Organizational Dynamics*, *34*(2), 141–152.

Jaques, T. (2007a). Issue management and crisis management: An integrated, non-linear, relational construct. *Public Relations Review*, *33*(2), 147–157.

———. (2007b). Issue or problem? Managing the difference and averting crises. *Journal of Business Strategy*, *28*(6), 25–28.

———. (2008). A case study approach to issue and crisis management: Schadenfreude or an opportunity to improve? *Journal of Communication Management*, *12*(3),192–203.

———. (2009a). Integrating issue management and strategic planning: Unfulfilled promise or future opportunity? *International Journal of Strategic Communication*, *3*(1), 19–33.

———. (2009b). Issue management as a post-crisis discipline: Identifying and responding to issue impacts beyond the crisis. *Journal of Public Affairs*, *9*(1), 35–44.

———. (2009c). Learning from past crises: Do iconic cases help or hinder? *Public Relations Journal*, 3(1). Retrieved from http://www.prsa.org/prjournal/

Kash, T. J., & Darling, J. R. (1998). Crisis management: Prevention, diagnosis and intervention. *Leadership & Organization Development Journal*, *19*(4), 179–186.

Kovoor-Misra, S. (1995). A multi-dimensional approach to crisis management for technical organizations: Some critical factors. *Technological Forecasting and Social Change*, *48*(2), 143–160.

Kovoor-Misra, S., Zammuto, R. F., & Mitroff, I. I. (2000). Crisis preparation in organizations: Prescription versus reality. *Technological Forecasting and Social Change*, *63*(1), 43–62.

Larkin, J. (2003). *Strategic reputation risk management*. London: Palgrave Macmillan.

Lauzen, M. M. (1995). Toward a model of environmental scanning. *Journal of Public Relations Research*, *7*(3), 187–203.

Lipton, E., & Shane, S. (2005, September 4). Homeland security chief defends federal response. *The New York Times*. Retrieved from http://www.nytimes.com/2005/09/04/national/nationalspecial/04fema.html

Meng, M. (1992). Early identification aids issue management. *Public Relations Journal, 48*(3), 22–24.

Mitroff, I. I. (2001). *Managing crises before they happen*. New York: AMACOM.

———. (2002). Crisis learning: The lessons of failure. *The Futurist, 36*(5), 19–21.

———. (2004). *Crisis leadership: Planning for the unthinkable*. Hoboken, NJ: John Wiley.

Mitroff, I. I., Harrington, K. L., & Gai, E. (1996). Thinking about the unthinkable. *Across the Board, 33*(8), 44–48.

Mitroff, I. I., & Pauchant, T. C. (1990). *We're so big and powerful nothing bad can happen to us: An investigation of America's crisis-prone corporations*. New York: Birch Lane.

Nystrom, P. C., & Starbuck, W. H. (1984). To avoid organizational crises, unlearn. *Organizational Dynamics, 12*(4), 53–65.

Pauchant, T. C. & Mitroff, I. I. (1992). *Transforming the crisis-prone organization: Preventing individual, organizational and environmental tragedies*. San Francisco: Jossey-Bass.

Pearson, C. M., & Mitroff, I. I. (1993). From crisis prone to crisis prepared: A framework for crisis management. *Academy of Management Executive, 7*(1), 48–59.

Penrose, J. M. (2000). The role of perception in crisis planning. *Public Relations Review, 26*(2), 155–171.

Phelps, R. (2005, September 21). All hazards vs homeland security planning. *Disaster Resource Guide*. Retrieved from http://www.ems-solutionsinc.com/pdfs/MeettheExperts.pdf

Renfro, W. (1993). *Issues management in strategic planning*. Westport, CT: Quorum Books.

Roux-Dufort, C. (2000). Why organizations don't learn from crises: The perverse power of normalization. *Review of Business, 21*(3/4), 25–30.

———. (2007). A passion for imperfections: Revisiting crisis management. In C. M. Pearson, C. Roux-Dufort, & J. A. Clair (Eds.), *International handbook of organizational crisis management* (pp. 221–252). Thousand Oaks, CA: Sage.

Select Bipartisan Committee. (2006). *A failure of initiative: Final report of the select bipartisan committee to investigate the preparation for and response to Hurricane Katrina*. Washington, D.C.: US Government Printing Office.

Shrivastava, P. (1988). Industrial crisis management: Learning from organizational failures. *Journal of Management Studies, 25*(4), 283–284.

———. (1995, August). Ecocentric management for a globally changing crisis society. Paper presented at National Conference, Academy of Management. Vancouver, B.C., Canada.

Smith, D. (1990). Beyond contingency planning: Towards a model of crisis management. *Industrial Crisis Quarterly, 4*(4), 1–26.

———. (2005). Business (not) as usual: Crisis management, service recovery and the vulnerability of organizations. *Journal of Services Marketing, 19*(5), 309–320.

Starbuck, W. H., & Milliken, F. J. (2007). Challenger: Fine-tuning the odds until something breaks. *Journal of Management Studies, 25*(4), 319–340.

Stern, E. (1997). Crisis and learning: A conceptual balance sheet. *Journal of Contingencies and Crisis Management, 5*(2), 69–86.

Stocker, K. P. (1997). Strategic approach to crisis management. In C. L. Caywood (Ed.), *The handbook of strategic public relations and integrated communications* (pp. 189–203). New York: McGraw-Hill.

Stout, D. (2009, September 3). Report details how Madoff's web ensnared SEC. *The New York Times*, pp. B1.

Swartz, E., Elliott, D., & Herbane, B. (1995). Out of sight, out of mind: The limitations of traditional information systems planning. *Facilities*, *13*(9/10), 15–21.

't Hart, P., Heyse, L., & Boin, A. (2001). New trends in crisis management and crisis management research: Setting the agenda. *Journal of Contingencies and Crisis Management*, *9*(4), 181–188.

Turner, B. A. (1976). The organizational and interorganizational development of disasters. *Administrative Science Quarterly*, *21*(3), 378–397.

Vidal, J. (1999, October 7). We forgot to listen, says Monsanto. *The Guardian*. Retrieved from http://www.guardian.co.uk/science/1999/oct/07/gm.food

Wartick, S. L., & Rude, R. E. (1986). Issues management: Corporate fad or corporate function? *California Management Review*, *29*(1), 124–140.

Chapter 3

Participatory Pre-crisis and Crisis Communication: A Conceptual Approach

Jesper Falkheimer & Mats Heide

This chapter challenges readers to re-conceptualize previous thinking of crisis and crisis management. They see today's crisis as one that is less static in nature—an event that does not cross spatial and functional boundaries. In essence, the crisis does not necessarily have the definitive beginning and end that much of current thinking about crisis would seem to suggest. This chapter proposes a threefold "system shift" for understanding crisis. First, the focus should shift from post-crisis (where it has long existed) to pre-crisis. Second, scholars and practitioners should break the boundaries that exist between internal and external communication. Finally, crisis response should incorporate a more participant-oriented approach, which necessitates a move away from reliance on media and toward understanding more about how meaning is created in society.

Introduction

Crisis communication constitutes one of several different forms of strategic communication between organizations, publics, and society. This means, of course, that when structural conditions for communication in society change, crisis communication must adapt to such changes. Today we live in a global society in which economic, cultural, and political systems cannot be isolated to national contexts. Crises spread rapidly over time and place and encompass many different actors. This is a development that is ultimately a consequence of modernity. However, recommended strategies still often deviate from traditional crisis management even today—not in accordance with the changes in society that have taken place during past decades. In contemporary society the possibilities of controlling communication processes have become more complex, and they demand different parallel strategies depending on context, situation, and culture. Hence, there is no one best way anymore. Pre-crisis presents the greatest scope for action management with regard to building confidence, flexible communication systems, and action preparedness. Consequently, pre-crisis communication is more important than ever before.

From a communicative perspective, the present constitutes a system shift: from that of a more solid society based on mass communication, centralization, and clear boarders to that of a liquid society based on a high level of complexity and new forms of participation from different groups. Contemporary society has been analyzed by divergent social scholars and is characterized as a late modern risk society, network society, or just liquid society (Bauman, 2000), which is the term we will use in this chapter. Common characteristic features of a liquid society are insecurity, fragmentation, individualism, and increased reflexivity—concepts with clear connections to and great relevance for crisis communication. This is the point of departure of the chapter, one which aims to integrate a number of concepts and modes of thought rooted in current social theory as well as media and organization research, of potential value to crisis research.

The bulk of crisis research has arisen from a traditional informational perspective based in solid society. Four problems underline this research: (1) how citizens or other interested stakeholders comprehend crisis, (2) how such stakeholders make use of crisis information, (3) the role and significance of news media, and (4) which sources dominate the media's reporting (Tierney, Lindell, & Perry, 2001).

Within research pertaining to organizational aims, two orientations are evident (Seeger, Sellnow, & Ulmer, 2001). Researchers have been primarily interested in different defense strategies or rhetorical strategies for impression management, which is considered to lead to the improvement of an organization's image or situation either during or in the aftermath of the occurrence of a crisis (e.g., Benoit, 1995). However, there is also a large research area that is highly interested in the phase prior to the occurrence of a crisis, with a particular focus on risk analysis and action programs designed to prevent diverse issues that lead to crisis. This area (issue management) has less interest in out-and-out communication issues; instead, it concerns itself primarily with management and opinion formation from an instrumental perspective.

Collectively, traditional crisis communication theories analyze crisis as a clear consequence of a single issue, view crisis as a linear transmission process (with a beginning and an end), and have a marginal interest in how different publics or stakeholders interpret messages and processes

(Falkheimer & Heide, 2009). New research in crisis management opposes the traditional view on crisis. For example, Boin and Rhinard (2008) use the concept of *transboundary threats* to understand threats that cross spatial as well as functional boundaries and are not possible to view as a causal consequence of a simple issue.

We claim that the system shift presents three new demands, which we will explain and argue for in this chapter. The first demand is that the focus for both researchers and practitioners, without neglecting operative emergency crisis management, is transferred to pre-crisis as well as post-crisis. This entails that the line of demarcation between crisis and risk communication becomes blurred. Crisis communication research cannot merely emanate from a meso-perspective (organizational level) but must also be firmly established partly in a macro perspective (social theory) and partly in micro theory (people's interpretations and actions in diverse situations and contexts).

The second demand is that the boundaries between internal and external communication are questioned. Crisis communication has previously focused primarily on the latter dimension, neglecting the links between internal and external communication. The source of this over-passing of boundaries is brought about by the contemporary development of organizations where increased transparency, employees' communication, and open organization forms challenge traditional approaches to organizations' communication. The area of internal crisis communication has received very little attention within research and crisis management (see Sellnow, Veil, & Streifel, 2010), which is surprising considering internal crisis communication is of fundamental importance. Organizations' responses to crises are a direct result of the complex internal processes and communication among employees (cf. Boin, 't Hart, Stern, & Sundelius, 2005). In order for this research area to develop, researchers need to focus on organizations' internal processes, relations, and communications to a great extent (Taylor, 2010).

A third demand is to adopt a participant-oriented approach to crisis communication, one that has its point of departure in the media and communication development of society. Here, the emphasis is (to a great extent) to develop a more culture-oriented perspective regarding crisis communication, thereby placing more accentuation on meaning-generating and active participation. In its entirety, this entails that organizations adopt a holistic perspective regarding crisis communication, which necessitates the challenging of intra-organizational, economic, social, and cultural boundaries.

In the first section of this conceptual chapter, which includes a number of examples to illustrate our line of reasoning, we discuss the evolving societal conditions and participatory culture. In the second section, we further develop the organizational aspect from a social constructionist perspective and develop the meaning-generating perspective attitude that is crucial for a participant-oriented approach. In conjunction with this, we then, in the concluding section, focus on co-workership and its significance in order to emphasize possible consequences that can develop pre-crisis communication as an area of research and practical experience. The premise here is that strategies must be seen as practical experiences, which are influenced by both societal conditions and people's meaning-making and which, in turn, influences these conditions and processes. That is, strategies cannot be isolated on an organizational level (Hallahan, Holtzhausen, van Ruler, Verčič, & Sriramesh, 2007).

Crisis Communication in a Liquid Society

An article on strategic communication as a concept and on its various sub-disciplines observes that:

> It is important to note that these disciplines were developed as specialty functions in the modernistic world of the 20th century. Yet, at the beginning of the 21st century, these disciplines function in a postmodern environment that stresses more holistic approaches to examining organizational phenomena, while having to deal with increasingly fragmented audiences and delivery platforms. (Hallahan, Holtzhausen, van Ruler, Verčič, & Sriramesh, 2007, p. 4)

Hence, crisis communication should no longer be isolated from other forms of strategic communication. If one defines crises from a comprehensive perspective—occurrences that create great uncertainty for citizens and interested parties, threaten the values and legitimacy of organizations or society, and necessitate quick decisions made on shaky grounds (Boin, ′t Hart, Stern, & Sundelius, 2005)—one can also advocate that crises are, if anything, continuous processes rather than temporary anomalies in normal systems. This indicates that crisis communication is transported closer to other fields of strategic communication as risk communication and internal organizational communication.

The system shift that alters the conditions for crisis communication has been described in a variety of ways. The sociologist Zygmunt Bauman (2000) differentiates between solid society and liquid society, the former being characterized by a centralization of diverse institutions and by clear roles for different actors in society. In this societal structure, public authorities, citizens, and customers have recognizable boundaries separating them. Mass communication constitutes the dominant form of communication, and people define themselves through distinguishable groups. Further, the conveyance of messages using a select few mass media outlets to inform the masses is the given strategy. Liquid society, which develops parallel to solid society and, indeed, has become all the more a dominant social system, is defined by increasing insecurity and more individual choices. Individuals experience increased uncertainty and continually re-examine their relationships in accordance with how they interpret new information. The boundaries, which existed in a solid society, are constantly challenged. The traditional transmission model of communication, which is founded on a one-way mode of communication, is no longer applicable. In a liquid society, everyone is a sender, or rather, a participant.

Media researcher James Lull (2007) describes the development in a somewhat similar manner as Bauman (2000). He claims that we now live in a world where two conflicting media cultures collide: cultures based on different modes of communication. On the one hand, we have traditional *push* media cultures, where centralized authorities and institutions deliver professional mass media content through controlled one-to-many channels. Audiences are apprehended as target-groups, defined through demographic variables. Journalists and professional communicators are the senders while citizens, customers, or publics are the recipients. During a crisis the main advice for effective communication, according to this logic, would be to quickly build a centralized and tight system in order to gain control over message formulation and transmission. The solid society and the media culture of push belong together. Lull (2007) contrasts the push culture with pull culture, which is defined as a media culture where the needs and interests of niche publics or individual media consumers are in focus while old sender-recipient-models are made obsolete. While *push* culture is collective, slow-paced, closed, community-bound, uniform, and based on a paradigm of production, *pull* culture is individual, private, fast-paced, micro-oriented, open, diverse, fragmented, and based on a paradigm of active consumption. Both cultures exist in contemporary society, even though *pull* culture obviously is becoming more important, especially among younger generations.

Jenkins (2006) believes that we live in a convergence culture in which old and new media integrate in society: "If the digital media revolution paradigm presumed that new media would displace old media, the emerging convergence paradigm assumes that old and new media will interact in even more complex ways" (p. 6). Contemporary crises have a lot in common with the concept of convergence, which Jenkins (2006) defines as "the flow of content across multiple media platforms, the cooperation between multiple media industries, and the migratory behavior of media audiences who will go almost anywhere [...]" (p. 2). Media convergence interacts with participatory culture as a central concept and discourse. The participatory culture is linked to active consumption and collective processes, primarily between participant–participant rather than between organizations–participants. The ideas behind participatory culture are not new but have a long history. There are links to the pragmatist communication theory developed a century ago as well as the critical theory of communicative and strategic communicative action developed by Habermas (1984). But participatory theory now has a renaissance linked to the consequences and potential of new and social media, which enforces the development.

From Sender Perspective to Participant Perspective

Besides participatory culture, the other major trait that Jenkins (2006) discusses is collective intelligence, based on the fact that "[...] none of us can know everything; each of us knows something; and we can put the pieces together if we pool our resources and combine our skills" (p. 4). A typical example of this is when media users interact in different social media, asking questions and getting answers from other users, while trying to understand risks or crisis situations. A typical arena for collective intelligence—and crisis sense-making—is the online community, image, and video hosting website Flickr, launched in 2004 ("Share your photos. Watch the World" is the catchphrase, see www.flickr.com). The website is used by many bloggers and hosts millions of images and videos as well as groups connected to different issues. For example, during and after the earthquake in Chile on February 26, 2010, thousands of pictures and videos were shared on the website, and discussion groups developed new information on what happened. In corporate crises, activism also is quickly organized through social media, creating collective intelligence as well as opinion formation and forcing authorities and corporations to act. For instance, the Dole Food Company was criticized heavily in the Swedish documentary movie *Bananas* (2009), which focused on the labor conditions in banana plants. Dole sued the filmmaker, but the summons led to national and international debate, and communities gathered online, among them a *Bananas* support group on Facebook with thousands of supporters. Finally, Dole decided to drop the charges. The film probably got a lot more attention, partly because of activism on the Internet, than it would have received if Dole would have chosen another strategy.

Media corporations and other actors no longer control communication processes; instead, these take on a more chaotic expression. Consequently, information that is conveyed from an organization to a customer or citizen is just one of several processes and is unlikely to be that of greatest importance with regard to interpretation and action. More corporations and authorities are experiencing this with greater frequency. An example is when the computer corporation Creative—renowned specialists in sound cards, particularly its sound card series Soundblaster—found itself in a crisis of confidence. The problem was that many of its sound cards became useless when Microsoft launched its new operating system, Windows Vista 2007. In response, Creative promised on its Internet forum to find a solution directly. However, this did not materialize, and 18 months later the corporation was still maintaining that it would find a solution to the problem. The matter was resolved when a forum member suddenly managed to come up with an answer to the complication. His solution spread quickly through social media. However, Creative's response was far from appreciative. Instead, Creative threatened to take legal action if he failed to withdraw his solution, which the corporation considered an infringement on its corporate rights. This denunciation was published on the Internet by the forum member in question, which lead to widespread mistrust of Creative by its users and their challenge of Creative's position on the sound card issue (Wired Magazine, 2008).

In a participatory culture, actors and co-producers forward their views through various forms of media. Jenkins's discussion on participant culture describes how oppositional readings of various messages or events lead to action as well as how corporations utilize participant culture in order to reach their particular goals. It should be noted that the theory pertaining to participatory culture has its weaknesses. Power, knowledge, and resources have significance, and some actors have greater opportunities to influence than others. Crises situations are seldom characterized by a desire to accommodate participant culture; instead, the goal is to quickly and effectively reach out with a concise message, which would lead to a certain action. However, at the same time, all organizations take into consideration the possibilities that participatory culture offers diverse actors—from interested parties to individuals—to create their own forums and narratives around causes and consequences of crises. The most interesting strategic question is how can a participant-oriented pre-crisis perspective contribute to the building of action-competence? From a crisis communication perspective, the development of pull culture as well as participatory and convergence culture in a liquid society challenges the guidelines usually given to different actors. The organizational centralization of crisis information and the willingness to provide a clear and harmonized message—the usual recommendation—can appear useless in a societal and organizational context, which is characterized by increasing decentralization and user-orientation. The desire for consistency and control that dominates crisis communication, as well as in strategic communication, must be balanced against the contemporary communication structure that is emerging.

From a Modernist to a Late-Modern Organizational Approach

The modernist perspective on organizations—in which objective and stable phenomena have been reviewed—has been dominant for some time. The premise is that management has full control, information, and first-rate knowledge; as a consequence, it is regarded as being best equipped to make rational decisions for an organization. Furthermore, it is also assumed that leaders can control events in their surroundings. This perspective is based on an acute sense of confidence in rationality in systems at all levels, which stems from the modernist assumptions that prevailed in the United States in the 1960s (Whittington, 2004). Accordingly, man is viewed as rational according to the logic of Adam Smith's *Economic Man*, and causes and effects may be controlled. To obtain optimal effects, organizational executives (who have very important roles here) must create planning functions, which influence all decisions. Unquestionably, the modernist perspective has had a most notable effect on the field of crisis communication. Traditionally, crisis management and crisis communication have been perceived as crash management. That is, they concern themselves with quenching conflicts that suddenly flare up as well as damage control. In short, the emphasis is to reduce damage (Johansen & Frandsen, 2007). From this perspective, it is management who identifies a crisis and initiates activities to administer the crisis. Consequently, our premise is that the boundary-line between risk and crisis is not totally transparent. However, it is clear that someone within an organization must observe and recognize that a crisis is looming. Crisis, we claim, is about strategic choice. However, the quality of the thinking, which precedes such a choice, is critical with regard to how a crisis is dealt with. The primary problem is that policymakers—in practice, often managers—are so focused on strategic plans and ideas that they rarely take a step back to reflect on strategic choices (Gavetti & Rivkin, 2005). When an organization faces a new threat or new opportunity, decision-makers turn to past experience and apply the lessons learned there to the new situation.

A typical trait of the modernist perspective is that the environment is perceived as objective—that is, something that exists "out there." Environments that are often thought of in terms of organizational literature have certain characteristics, such as a set of components (e.g., economic, technological, and social); interested parties (e.g., customers, politicians, and suppliers); and as a set of attributes (e.g., instability and complexity) (Sutcliffe, 2001). When an environment is presumed to be objective, it is assumed that organizations are embedded in external, independent environments that have different powers, which they must attempt to control and adapt to (Smircich & Stubbart, 1985). It is believed that it is possible to measure and determine an area, and the organization's nourishment is to meet and react to the environment's real obstacles and challenges.

Communication is seen here as a tool to provide and spread information (Varey, 2000). The focus is on which formulation of the customized message and choice of media is the most efficient method to reach the receiver. However, it is up for discussion as to how effective this type of communication is, as all people interpret information differently depending on their particular experience, interests, background, knowledge, and values. Communication is seen as a process of information transmission from A to B. However, what is regarded as problematic is the transmission itself—the transmission of information. Effective communication is deemed to have been achieved when the recipient has obtained the information. The interpretation aspect, however, is completely ignored in favor of a focus on selecting the right form of media for the right audience and for the right purpose.

In a late-modern perspective, it is assumed that there is no objective reality given in advance (Falkheimer & Heide, 2006). Rather, it is assumed that people create the environment, which in turn places limitations before us. The creation of environments occurs through communication between organization members. Hence, it is not only about members interpreting objective information that is "out there" but also that they interpret the information and then create a socially constructed reality. This means that a situation becomes a crisis when one or more interested parties perceive it as such (Coombs, 2007; Fearn-Banks, 2007; Gilpin & Murphy, 2008).

The American organizational psychologist Karl E. Weick (1969, 1979, 1995, 2009) has been of great importance to the development of the process-oriented view of organizations. Weick began to write in the 1960s when other organizational researchers were confused over the changing trends, which he tried to describe and explain using traditional models and concepts. Weick has primarily focused on sense-making, the process by which people try to create meaning in ambiguous situations and react in different ways in order to reduce this ambiguity. According to Weick, some organizations are not stable entities with objective characteristics. Instead, he speaks of organizations as clusters of social relations, which are never stable. Accordingly, continual organizing processes occur, which are activities that offer more order in socially constructed reality by reducing ambiguity. Thus, an organization is about creating order in a chaotic world and channeling it toward a certain goal with the help of particular significant frameworks and rules. An important part of the organization is sense-making, which occurs when organization members fail to comprehend a situation. Hence, sense-making occurs when we find ourselves in a confusing or unclear situation. We are constantly told that we live in a changing and complex world where a great deal of information is exchanged. In order for us to act upon this sheer volume and complexity of information, we try to focus on a certain section of the information and ascribe some sense of meaning to it. From this, we create some understanding, and a socially constructed reality arises. Consequently, the socially constructed reality is expressed when organization members communicate and act jointly; this pattern of texts, verbal communication, and conduct forms the organization's everyday life (Weick, 2006).

Weick (1995) stresses there is a close link between organizations and communication: "If the communication activity stops, the organization disappears. If the communication activity becomes confused, the organization begins to malfunction. These outcomes are unsurprising because the communication activity is the organization" (p. 75).

In the environment, there are endless numbers of threats and opportunities. Policymakers highlight certain aspects of information that prevail. Once certain information in the environment has become stimuli—that is, it is deemed important and relevant—policymakers proceed to give the information a sense of meaning so it becomes meaningful (Sutcliffe, 2001). A key concern is that people act according to their understanding and interpretation, and that understanding is created with others (Maitlis, 2005). Formal organization and formal communication, such as leadership, intranet, and staff meetings, are important for collaborators' comprehension. However, the importance of informal networks is often neglected. A large proportion of people's interpretation of diverse events, the dissemination of information and production, and the sharing of knowledge occur via informal networks (Weick, 2009). Informal networks are therefore fundamental to the crisis management of organizations, which a great deal of research has failed to highlight.

Weick (1995) points out that sense-making is retrospective. It is only after we have acted that we understand what we have been through. Sense-making is not about finding truth or acting correctly but about people attempting to answer the question: "What is going on?"

This question arises in all new situations in which we do not know what to do. It is only after we have acted or spoken that we understand what has happened. According to Weick (1995, p.12), the method for sense-making is, "How can I know what I think until I see what I say?" Indeed, many thinkers before Weick have observed this relationship. Already in the late 1800s, the Danish philosopher Søren Kierkegaard pointed out that as life can only be understood later, it must be lived forward. What Weick has contributed is the connection between understanding, action, and communication.

The disadvantage with Weick's "How can I know what I think until I see what I say?" is that it focuses solely on the individual. Sense-making is, however, a social process that includes communication and collaboration with others. To create meaning within an organization means to structure and organize the unfamiliar, the unexpected, or the different in order to proceed to act upon this understanding.

In contrast to the modernist perspective, where it is assumed that leaders within an organization make final decisions and that they have full control, we claim that decisions emerge from various networks within the organization. Weick underlines that the risk with centralized decision-making is that the competence needed for problem solving is reduced. The real experts are placed throughout the organization and have great opportunities to detect emergency signals earlier and to act to prevent further development. Weick, Sutcliffe and Obstfeld (1999, p. 104) point out, "Many organizational settings contain a million accidents waiting to happen, but most organizations don't see things that way." On that basis, organizations have significantly greater chances of managing a crisis at an early stage by granting more authority to organization members and encouraging them to act on so-called weak signals. Early detection is obviously a prerequisite to preventing any change or event from developing into an extraordinary event or crisis.

A close link to sense-making is communication and action (Weick, Sutcliffe, & Obstfeld, 2005). Through collaboration, organization members can discuss, develop, and share understandings of a situation, which in turn lead to action. Communication and action do not occur as a linear sequence but in cycles. Sense-making can start when an organization member acts. For example, he or she may try a new method to solve a problem; thereafter, results are discussed with colleagues, and some understanding is created until a new solution is examined and new results arise. Sense-making can also begin when someone points out that something is amiss, which requires action in order to be understood. Weick et al. (2005) conclude that people create meaning through "acting thinkingly." Thus, while people interpret their knowledge in the familiar framework, they simultaneously have the courage to distrust it by testing new approaches and, in this way, arriving at new interpretations.

Meaning and understanding are created through persuasion, discussion, and influence (Sutcliffe, 2001). Fundamental to the interpretation process is that the information is interpreted within a given interpretive framework; only then does it become meaningful. This does not mean that truth is created—only a certain socially constructed understanding. What commonly occurs is that information is designated by some label or other. Among the most prevalent is *threat* or *possibility*, with regard to changes in the environment. Communication is crucial in order for the number of possible interpretations to be reduced and thus to enable coordinated action (Weick, 1995). After having reached a fairly common interpretation, policymakers can make a decision allowing them to act.

New Modes of Organizing: Co-workership

Within organization theory, the focus has primarily been directed at top management—for example, how such leaders interpret external events and how they try to influence organization members' interpretations through sense-giving (e.g., Gioia & Chittipeddi, 1991; Gioia & Thomas, 1996). There is also some research on middle management, particularly with regard to change processes (e.g., Balogun & Johnson, 2005). Although top and middle management are important within sensemaking processes, frontline employees, despite everything, are of greatest significance. The knowledge and the interpretive frameworks people have dictate what and for whom information is significant. This often is a problem in an organization, as its members interpret information in different ways depending on where in the organization they work, their background, their interests, and their knowledge. For instance, there are divergent interpretations of an organization's goals and future by people on various hierarchical levels (e.g., members of management and personnel in production), with dissimilar education backgrounds (e.g., economists and technicians), and different interested parties (e.g., union representatives and the CEO). Diverse groups focus differently, paying attention to and interpreting various aspects of an organization's environment. Conveying a large number of interpretations and confusion simultaneously makes it difficult for policymakers to grasp a situation (the socially constructed reality). Undoubtedly, the actors of greatest significance for successful crisis management and crisis communication are staff (Mercer & Kapcio, 2006). Staff is important before, during, and after a crisis. They are these organization members who, at an early stage, can detect weak signals of change outside and within an organization, interpret them, and act accordingly (cf. Weick & Sutcliffe, 2007), thereby avoiding or dampening the effects of a crisis. Uncertainties facing organization members are prominent features of crises. These include members wondering what is happening, what the organization is doing to find solutions, and what the crisis will mean for them (Millar & Heath, 2004; Ulmer, Sellnow, & Seeger, 2007). Weick (1988, p. 311) maintains that "people see those events they feel they have the capacity to do something about." Employees have direct contact with various audiences through documents and communication. Furthermore, employees produce new knowledge and strategies for action after a crisis and thus create better opportunities to avoid future crises (Falkheimer, Heide, & Larsson, 2009). Employees' recollections of past crises and the lessons learned from them are therefore important for how new crises are managed. The internal processes are consequently linked to the results of an organization's crisis response and may directly impact the surrounding community. Despite the importance of employees with regard to successful crisis management and crisis communication, research has almost exclusively focused on external publics, which, in various ways, are affected by crises.

During the last decade, the status of management and managers has decreased due to a derogation rooted in the nature of the work and the connection to bureaucracy and inflexibility (Brocklehurst, Grey, & Sturdy, 2010). At the same time, the status of co-workers has increased, which, among other things, is related to the expansion of post-bureaucratic and knowledge-intensive organizations (see Alvesson, 2004). These organizations—and very often their communication policies—are imbued with concepts such as teams, dialogue, workplace democracy, and employee engagement. These and other related concepts presuppose a power shift with new communication processes and concurrently also ascribe an active, not to say crucial, role to co-workers. An interesting concept in this context is co-workership. Conceptually, it is fairly new and has its roots in the

Scandinavian management style, which is usually goal oriented and focused on devolution, participation, and coaching. In the international literature, one finds the closely related concept of fellowship, which derives from recent leadership research where leadership is seen as a process of mutual influence (Tourish, Craig, & Amernic, 2010). As a concept, fellowship implies a dependent, passive, and submissive role, while co-workership also includes the relation to the organization at large and to colleagues (Andersson & Tengblad, 2009). Co-workership indicates a more active role for employees and can be associated with late-modern organizational theory with loosely coupled networks, teamwork, and an underlining of horizontal communication. We have not encountered the use of co-workership within crisis communication research, although some researchers have previously spoken of the need to see organization members as the real experts with the optimal conditions to deal with crises at an early stage.

Weick (1988) is positive to our ability to manage crises and believes that they are easier to monitor than what we usually think. If we have the idea that we can control, this will, according to Weick, lead us to note more factors we can influence; and by responding from outside of the conception, more complex information will be simplified.

Weick (1990) has conducted a thorough analysis of what went wrong with the airplane disaster at Tenerife in 1977, where two Boeing 747s—one from KLM (the Royal Dutch Airlines) and the other from Pan Am—collided, resulting in 583 fatalities. Among his conclusions, which have a bearing on all organizations, Weick underlines the need for members constantly to talk to each other. In organizations characterized by a closed-communication climate, misunderstandings and errors will not be discovered in time. Important advice to management is (1) be generous with information, (2) acknowledge the important function of information contained in the informal organization, and (3) do not wait too long to communicate bad news (Sutton & Kahn, 1987). Weick (1990, p. 589) argues that it is important to introduce a new standard in organizations: "If things do not make sense, speak up. Only by doing so can you break pluralistic ignorance (i.e., 'You too? I thought I was the only one who didn't know what was going on.')." Hence, it is through an open-communication climate that problems can be detected at an early stage. Moreover, open communication creates the best conditions for an organization in which the exchange of knowledge and experiences occurs between different units. Weick and Sutcliffe (2001) argue that the whole point of a learning organization is that it must find ways to deal with the fact that it does not know what it does not know. Should organization members report errors, it must be assumed that there is an open-communication climate and that management understands and appreciates information concerning mistakes. Only then can the organization as a whole continually learn and develop. A prerequisite for effective signal identification is the existence of shared meanings within an organization. These shared meanings can be both an asset and an obstacle. If the organization is overly confident and interprets events as it always has, it can be difficult to have the openness and curiosity required in order to interpret complex problems (Weick, 1996). Maitlis and Sonenschein (2010) postulate the need for two actions for shared meanings to have an adaptive rather than a destructive role with regard to the detection of crises. These are updating and doubting (Weick, 2010). Updating is an activity that allows organization members to reconsider interpretations based on new information. This, however, can be problematic if some *but not all* in an organization reconsider their interpretations. What is needed is information collection and dissemination of new knowledge that are based on collective knowledge. Doubt includes an understanding that all sensemaking is temporary and that it stimulates us to constantly seek to create new understanding. We can never conduct a definite interpretation of a situation because things are continually changing. A major risk for detection is when organization members believe they have reached an understanding and that they have found an answer (Maitlis & Sonenshein, 2010).

Internal Communication Strategies to Create Crisis Management Readiness

The majority of practitioners agree that organizations should have a clear profile and that all co-workers, in all situations, should support the organization's core values and comprehensive image. During crises, this requirement escalates: Internal consistency and composure are prerequisites of effective crisis communication. In practice, this is rarely optimal, but there are some key communication strategies that can be considered fundamental. Cheney (1983) has identified a number of communication strategies that integrate an organization, and we believe that these can also be linked to the developing of preparedness prior to crises.

The common ground theory focuses primarily on actions and communicative measures where the leader or the leadership of an organization demonstrates that management is related to all employees by sharing the background and being the driving force. In rhetoric, this strategy chiefly pertains to ethos—the sense of confidence employees have in their manager or management. Without this confidence, no leadership can function, especially under uncertain circumstances. In addition, confidence is difficult to achieve in situations in which it has fallen dramatically due to past events or crises. When a manager or management communicates without strong support, or in a manner that is perceived negatively by staff or the environment, a gap is created—"us and them"—that thereafter has the ability to reproduce again and again. Most importantly, the leader or leadership has to succeed in creating identification with all employees regarding the organization's core mission. Such a result can only be created through a long and well-established leadership balance between listening and decision-making. The worst-case scenario is when there is a vacuum in the organization, which increases uncertainty and rumors; consequently, motivation declines.

Identification of antithesis is another strategy to create better integration. This involves a conscious leadership strategy to create images of threat, competition, or a common enemy that results in all co-workers in an organization pulling together. Enemies are constructed consciously and can take several forms: individuals, groups, or organizations. Hostile images are projected from strategic management into the organization to increase internal cohesion. When crisis situations arise, such social processes appear. With a clear enemy image, it is easier to build consensus and rally around the organization's values. But this strategy may be viewed as unethical rather than constructive.

Unifying symbols is a third strategy that is deemed to integrate an organization and create a uniform brand externally and build pre-crisis reputation capital. These symbols constitute everything from logos, rituals, clothes, and architecture to personal brands of directors and visible employees. During crises, clear symbols that unify employees and communication with external publics are needed. These symbols are important for the interpretation of different messages and must ultimately be based on earning confidence. In the long term, confidence-building efforts are vital when an organization finds that crisis is eminent. This entails that crisis communication has to be seen as part of an organization's entire strategic communication. It is during a crisis that earned confidence and symbols are of greatest significance.

Conclusion

In this chapter, our primary aim has been to problematize crisis communication from a conceptual perspective. The line of argument presented supports the thesis that pre-crisis communication is central to the optimization of operational crisis communication and crisis management, which attracted greatest interest in previous research. Through the use of social theory, current research around media, participatory culture, and the internal communication of organizations, we believe that research on crisis communication can be developed further. What is missing in this chapter is partly empirical depth and partly concreteness as to which strategies practitioners can apply. We are aware of this, but the hope is that the theoretical and conceptual reasoning might inspire both researchers and practitioners to further the discussion. So far, crisis communication research has seldom questioned or problematized fundamental assumptions and communicative conditions.

References

Alvesson, M. (2004). *Knowledge work and knowledge-intensive firms*. Oxford: Oxford University Press.

Andersson, T., & Tengblad, S. (2009). Medledarskap: Ledarskap som kollektiv förmåga. In S. Jönsson & L. Strannegård (Eds.), *Ledarskapsboken* (pp. 245–268). Malmö: Liber.

Argenti, P. A. (2003). *Corporate communication (3rd ed.)*. Boston, MA: Irwin McGraw-Hill.

Balmer, J. M. T., & Gray, E. R. (2000). Corporate identity and corporate communications: Creating a competitive advantage. *Industrial and Commercial Training, 32*(7), 256–261.

Balogun, J., & Johnson, G. (2005). From intended strategies to unintended outcomes: The impact of change recipient sensemaking. *Organization Studies, 26*(11), 1573–1601.

Bananas. (2009). See http://www.bananasthemovie.com/

Bauman, Z. (2000). *Liquid modernity*. Cambridge: Polity Press.

Benoit, W. L. (1995). *Accounts, excuses, and apologies: A theory of image restoration strategies*. Albany, NY: State University of New York.

Boin, A., 't Hart, P., Stern, E. K., & Sundelius, B. (2005). *The politics of crisis management: Public leadership under pressure.* Cambridge: Cambridge University.

Boin, A., & Rhinard, M. (2008). The management of transboundary crises: What role for the EU? *International Studies Review, 10*(1), pp. 1–26.

Cheney, G. (1983). The rhetoric of identification and the study of organizational communication. *Quarterly Journal of Speech, 69*, 143.

Christensen, L. T. (2002). Corporate communication: The challenge of transparency. *Corporate Communications: An International Journal, 7*(3), 162–168.

Christensen, L. T., Morsing, M., & Cheney, G. (2008). *Corporate communications: Convention, complexity, and critique*. London: Sage.

Christensen, L. T., Torp, S., & Firat, A. F. (2005). Integrated marketing communication and postmodernity: An odd couple? *Corporate Communications: An International Journal, 10*(2), 156–167.

Coombs, W. T. (2007). *Ongoing crisis communication: Planning, managing, and responding* . Thousand Oaks, CA: Sage.

Falkheimer, J., & Heide, M. (2006). Multicultural crisis communication: Towards a social constructionist perspective. *Journal of Contingencies and Crisis Management, 14*(4), 180–189.

Falkheimer, J., & Heide, M. (2009). Crisis communication in a new world: Reaching multicultural publics through old and new media. *Nordicom Review*, 30, 55–65.

Falkheimer, J., Heide, M., & Larsson, L. (2009). *Kriskommunikation*. Malmö: Liber.

Fearn-Banks, K. (2007). *Crisis communication: A casebook approach.* Mahwah, NJ: Erlbaum.

Gavetti, G., & Rivkin, J. W. (2005). How strategists really think: Tapping the power of analogy. *Harvard Business Review, 83*(4), 54–63.

Gilpin, D. R., & Murphy, P. J. (2008). *Crisis communication in a complex world.* Oxford: Oxford University Press.

Gioia, D. A., & Chittipeddi. (1991). Sensemaking and sensegiving in strategic change initiation. *Strategic Management Journal, 12*, 433–448.

Gioia, D. A., & Thomas, J. B. (1996). Image, identity and issue interpretation: Sensemaking during strategic change in academia. *Administrative Science Quarterly, 41*, 370–403.

Goodman, M. B. (2000). Corporate communication: The American picture. *Corporate Communications: An International Journal, 5*(2), 69–74.

Habermas, J. (1984). *The theory of communicative action. Vol. 1: Reason and the rationalization of society.* Cambridge: Polity Press.

Hallahan, K., Holtzhausen, D., van Ruler, B., Verčič, D., & Sriramesh, K. (2007). Defining strategic communication. *International Journal of Strategic Communication, 1*(1), 3–35.

Hübner, H. (2007). *The communicating organization: Towards an alternative theory of corporate communication.* Heidelberg: Physica Verlag.

Jenkins, H. (2006). *Convergence culture: Where old and new media collide.* New York: New York University Press.

Johansen, W., & Frandsen, F. (2007). *Krisekommunikation: Når virksomhedens image og omdømme er truet.* Köpenhamn: Samfundslitteratur.

Llewellyn, N., & Harrison, A. (2006). Resisting corporate communications: Insights into folk linguistics. *Human Relations, 59*(4), 567–597.

Lull, J. (2007). *Culture-on-demand: Communication in a crisis world.* Malden, MA: Blackwell.

Maitlis, S. (2005). The social processes of organizational sensemaking. *Academy of Management Journal, 48*(1), 21–49.

Maitlis, S., & Sonenshein, S. (2010). Sensemaking in crisis and change: Inspiration and insights from Weick (1988). *Journal of Management Studies, 47*(3), 551–580.

Marken, G. A. (2001). Corporate communications: It's all about delivering value. *Public Relations Quarterly, 46*(1), 39–41.

Mercer, L., & Kapcio, P. (2006). Internal communications for the avian flu: Anticipating effects on lives and livelihoods. *Public Relations Tactics, 13*(7), 20–22.

Millar, D. P., & Heath, R. L. (2004). *Responding to crisis: A rhetorical approach to crisis communication.* Mahwah, NJ: Lawrence Erlbaum.

von Otter, C. (2006). *Inledning*: *Om ledarskap för medarbetarskap.* I C. von Otter (red.), *Ledarskap för fria medarbetare* (s. 11–26). Stockholm: Arbetslivsinstitutet.

van Riel, C. B. M. (1995). *Principles of corporate communication.* London: Prentice-Hall.

Seeger, M. W, Sellnow, T. L., & Ulmer, R. R. (2001). Public relations and crisis communication: Organizing and chaos. In R.L. Heath (Ed.), *Handbook of public relations* (pp. 155–165). Thousand Oaks, CA: Sage.

Sellnow, T. L., Veil, S. R., & Streifel, R. A. (2010). Credibility seeking through an interorganizational alliance: Instigating the Fen-Phen confrontation crisis. In W. T. Coombs & S. J. Holladay (Eds.), *The handbook of crisis communication* (pp. 657–674). Malden, MA: Wiley-Blackwell.

Smircich, L., & Stubbart, C. (1985). Strategic management in an enacted world. *Academy of Management Review, 10*(4), 724–736.

Steiner, C. J. (2001). How important is professionalism to corporate communication? *Corporate Communications, 6*(3), 150–156.

Sutcliffe, K. M. (2001). Organizational environments and organizational information processing. In F. M. Jablin & L. L. Putnam (Eds.), *The new handbook of organizational communication: Advances in theory, research, and methods* (pp. 197–230). Thousand Oaks, CA: Sage.

Sutton, R. I., & Kahn, R. L. (1987). Prediction, understanding, and control as antidotes to organizational stress. In J. W. Lorsch (Ed.), *Handbook of organizational behavior* (pp. 272–285). Englewood Cliffs, NJ: Prentice-Hall.

Taylor, M. (2010). Towards a holistic organizational approach to understanding crisis. In W. T. Coombs & S. J. Holladay (Eds.), *The handbook of crisis communication* (pp. 698–704). Malden, MA: Wiley-Blackwell.

Tengblad, S. (2006). Is there a "new managerial work?" A comparison with Henry Mintzberg's classic study 30 years later. *Journal of Management Studies, 43*(7), 1437–1461.

Tierney, K. J., Lindell, M. K., & Perry, R. W. (Eds.). (2001). *Facing the unexpected: Disaster prepardness and response in the United States.* Washington, DC: Joseph Henry.

Tourish, D., Craig, R., & Amernic, J. (2010). Transformational leadership education and agency perspectives in business school pedagogy: A marriage of inconvenience? *British Journal of Management, 21*, 40–59.

Ulmer, R. R., Sellnow, T. L., & Seeger, M. W. (2007). *Effective crisis communication: Moving from crisis to opportunity.* Thousand Oaks, CA: Sage.

Varey, R. J. (2000). A critical review of conceptions of communication evident in contemporary business and management literature. *Journal of Communication Management, 4*(4), 328–340.

Weick, K. E. (1969). *The social psychology of organizing.* Reading, MA: Addison-Wesley.

———. (1979). *The social psychology of organizing.* Reading, MA: Addison-Wesley.

———. (1988). Enacted sensemaking in crisis situations. *Journal of Management Studies, 25*, 305–317.

———. (1990). The vulnerable system: An analysis of the Tenerife air disaster. *Journal of Management, 16*(3), 571–593.

———. (1995). *Sensemaking in organizations.* Thousand Oaks, CA: Sage.

———. (1996). Prepare your organization to fight fires. *Harvard Business Review, 74*(3), 143–147.

———. (2006). Faith, evidence, and action: Better guesses in an unknowable world. *Organization Studies, 27*(11), 1723–1736.

———. (2009). *Making sense of the organization: The impermanent organization.* Chichester: John Wiley and Sons.

———. (2010). Reflections on enacted sensemaking in the Bhopal disaster. *Journal of Management Studies, 47*(3), 537–550.

Weick, K. E., & Sutcliffe, K. M. (2001). *Managing the unexpected: Assuring high performance in an age of complexity.* San Francisco, CA: Jossey-Bass.

———, & ———. (2007). *Managing the unexpected: Resilient performance in an age of uncertainty.* San Francisco, CA: Wiley.

Weick, K. E., Sutcliffe, K. M., & Obstfeld, D. (1999). Organizing for high reliability: Process of collective mindfulness. In R. I. Sutton & B. M. Staw (Eds.), *Research in organizational behavior, Vol. 21* (pp. 81–123). Stamford, CT: JAI.

———, ———, & ———. (2005). Organizing and the process of sensemaking. *Organization Science, 16*(4), 409–422.

Whittington, R. (2004). Strategy after modernism: Recovering practice. *European Management Review, 1*, 62–68.

Wired Magazine (2008). Daniel K., who fixed Creative′s broken Vista drivers, speaks out (www.wired.com/gadgetlab/2008/04/daniel_k-who-fi/)

Chapter 4

The Ethical Challenges of Pre-crisis Communication

Shannon A. Bowen

This author provides valuable work that extends the anticipatory model to include a much greater recognition of ethical concern and analysis. The author begins by addressing the intersection between issue management, crisis management, and ethics. The overlapping of these concepts as well the complementary nature within them is explored to help establish the need for anticipatory ethical analysis. The author then details a seven-step process for expanding the anticipatory model to result in ethically conscious pre-crisis planning.

Introduction

This chapter focuses on the importance of ethics and anticipatory ethical analyses in the pre-crisis phase as a way to help prevent and minimize the impact of crises on an organization. A well-conceptualized and planned program of ethical decision-making can help an organization identify issues before they become crises, thereby effectively conducting pre-crisis management.

The key questions herein are: *what* needs to be communicated and *how* should it be communicated prior to a crisis to make the organization ethical? Successful issues management should help to prevent a crisis, but it should also be carried out from a foundation of moral responsibility using both utilitarian and deontological analyses. The ethics of pre-crisis planning are discussed through the use of the adapted and revised steps in issues management to include an ethical decision-making framework for analysis in each step.

Conceptualization

In the tripartite structure of pre-crisis, crisis, and post-crisis already discussed in this volume and other resources (Coombs, 1999), there is little doubt that ethical consideration fits squarely into the pre-crisis planning of any organization, allowing ethical analyses to also be implemented in the subsequent phases. In Olaniran and William's (2001, 2008) anticipatory model of crisis management, the anticipation of potential crises based on technological or human failures allows managers to build scenarios based on crisis potential and to lower the risk of a crisis actually occurring. In addition to technological or human failures, as described in the anticipatory model, there is no doubt that a thorough analysis of ethical implications and implementation of an ethical management plan can prevent many crises from occurring.

Ethics also plays an important role not only in anticipating crises and pre-crisis planning but also in the next phases of crisis and post-crisis recovery. Conducting analyses of responsibility can allow an organization to respond ethically to competing duties in the crisis phase and to evaluate that a moral intention was maintained and perceived by affected publics post-crisis. Current research on crisis management involves concepts that can all be traced back to a thorough ethical analysis by the organization, such as Heath, Lee, and Ni's (2009) study of community involvement and how crisis information is delivered; Avery and Kim's (2009) discussion of consistency as an important variable in crises; Jin and Cameron's (2007) study of emotional responses of public relations professionals (as opposed to conducting rational moral analyses); Huang's (2008) discussion of the importance of consistency in crises for building trust with publics; and Olaniran and William's (2008) study of organizations who evade their ethical responsibility through tactics such as scapegoating or denial. Each of the concepts studied by these scholars in relation to crisis involves an ethical component that can be revealed through moral analyses, such as those offered in this chapter.

Issues

As an initial and essential component of pre-crisis management, an organization should conduct a thorough *issues management* program in order to identify potential crises. There is certainly some blurring of the lines between issues management and crisis management (Heath & Palenchar, 2009; Jaques, 2009; Lauzen, 1997), but in the pre-crisis phase, the two concepts are inextricably linked through the priority that both place on the identification of issues. What are issues? Potential crises are also known as issues and were defined by an early leader in the field, Chase (1984), as “an unsettled matter which is ready for decision” (p. 38). Grunig and Repper (1992) took a more consequentialist-based approach in discussing issues as originating from the consequences an organization’s business has on publics.

Heath (1997), building on his earlier definition with Nelson (Heath & Nelson, 1986), defined an issue as “a contestable question of fact, value, or policy that affects how stakeholders grant or withhold support and seek changes through public policy” (p. 44). As you can see, Heath’s (1997) definition places an issue firmly in the public policy arena. That placement implies an ethical responsibility to do not only what is in the interest of the organization but also what is in the interest of multiple publics and ultimately in the public’s interest. For this reason, ethics and defining issues are inextricably entwined.

Issues Management

Issues management is the process of identifying potentially problematic issues and addressing those issues to organizational policy before they become crises. Although defining issues management is contentious (Gaunt & Ollenburger, 1995), it is definitely defined as a management function (Lauzen & Dozier, 1994) that is absolutely essential in the prevention of crises. Although Heath (1997) defined crisis management as a function of the larger process of issues management, Jaques (2007) summarized the relationship by concluding that the two functions are “intimately linked” (p. 285). Whether one sees pre-crisis planning (Coombs, 1999) as part of issues management (Heath, 1997) or views crisis preparedness separately (Jaques, 2007), it is clear that issues management should be understood and involved as an integral part of the process. Heath (1997) explained the overarching nature of the function:

> Issues management includes the identification, monitoring, and analysis of trends in key publics’ opinions that can mature into public policy and regulatory or legislative

> constraint of the private sector…the underpinning principle of issues management is not to avoid legislation or regulation but to balance the interests of all segments of the community so that each enjoys the proper amount of reward or benefit in proportion to the cost of allowing industry free rein to impose its own operating standards. (p. 6)

Issues management scholars have long discussed the anticipation of potential problems, changes, or threats. In clarifying the goals of issues management, Renfro's (1993) book elucidated: "The overriding goal in issues management function is to enhance the current and long-term performance and standing of the corporation by anticipating change, promoting opportunities, and avoiding or mitigating threats…" (p. 107).

When the issues manager defines an issue, it may or may not involve a pre-crisis situation (Coates, Coates, Jarratt, & Heinz, 1986). Because of the rapidly unfolding nature of issues and near infinite bifurcations (Murphy, 1996) and possibilities that emerge from the competing, multiple interests involved in issues, it is almost impossible to predict what issues have the potential to become crises and what issues will be of a more manageable nature. Certain issues lend themselves more naturally to evolving into a crisis (Coates et al., 1986; Coombs, 2007), such as potential labor strikes, natural disasters, unsafe working conditions, ethically questionable actions, and so forth. The ethical implications of most issues and the increasing engagement of publics using the Internet and social media have created higher demands for transparency and moral accountability than ever before. Even issues of a seemingly more mundane nature are encountered on a regular basis by issues managers, but they often can and do evolve into crises—especially when not managed properly (Mitroff, 2003) or when they involve an issue that a public identifies as an ethical value.

For example, the seemingly mundane Twitter post of the Ketchum public relations executive about how he did not like the home city of a major client, FedEx, quickly evolved into a crisis because Ketchum did not manage the issue. Once FedEx learned of the derogatory Twitter post, they publicly condemned the Ketchum agency and questioned both the ethical responsibility of the executive and Ketchum's credibility as experts with social media usage. Such a damaging situation for both Ketchum and the client could have been avoided through the *proactive* use of issues management. That process would have immediately identified the post as a problematic issue, asked the executive to delete it, would have sought to build and repair relationships with FedEx; likewise, a longer-term organizational policy for the use of Twitter by executives acting in a business context would have been established. In fact, that type of morally responsible and proactive issues management on the part of Ketchum could have served to strengthen relationships with FedEx by enhancing mutual interests, commitment, trust, and relationship satisfaction (Huang, 2001). It is unfortunate that such relationship enhancement did not occur, but we can learn much from the bungled handling of this issue that became a crisis for Ketchum (ironically, an agency that provides issues and crisis management services).

Clearly, issues must be handled with ethical tenets and moral responsibility in mind (Bowen, 2004b; Goldberg, 1993; Sims, 1994). The responsibility of the issues manager is not only to determine how to manage the issue or potential crisis but also to determine and research related facts. In defining the issue, the issues manager has a great deal of responsibility to decide what information is relevant, to decide who needs the information, and to determine when research is warranted and the form that research will take. The ethical challenges are immense.

Defining an issue that could potentially be a crisis requires perhaps the most moral discernment of all the conundrums faced by communication professionals. It requires a high level of discernment because if key facts are omitted, an ethical violation of the publics' right to relevant information has already been committed—even before any decisions are made on how to handle the issue. Further, those publics might perceive a misjudgment by the issues manager as an intentional concealment on the behalf of the organization, or worse, as an intentional deception.

A Conceptual Guide to Issues Management

The "public issues management system" as designed by Buchholz, Evans, and Wagley (1994, p. 41) provides a solid conceptual guide for managing issues in order to prevent those issues from turning into crises. However, their system does not recognize the importance of *ethical analyses* in issues management, nor does it recommend specific ethical tests or guidelines; rather, the authors *assume* an ethical basis but do not include it in their model. This chapter seeks to integrate the framework posed by Buchholz, Evans, and Wagley (1994) with current thought and research from moral philosophy and ethics.

The "public issues management system" offered by Buchholz, Evans, and Wagley (1994) includes six steps:

1. Identify public issues and trends in public expectations.
2. Evaluate their impact and set priorities.
3. Conduct research and analysis.

4. Develop strategy.
5. Implement strategy.
6. Evaluate strategy. (p. 41)

In order to adapt their issues management steps for pre-crisis planning and communication usage, a re-organization and the addition of a seventh step is necessary to explicitly include the evaluation of ethics, as explained below. The revised steps below also include a greater emphasis on the anticipatory model (Olaniran & Williams, 2001) for use in scenario buildings to create greater understanding of potential crises. The revised or adapted steps for ethically conscious pre-crisis planning and communication are:

1. Monitor the environment to identify public issues and trends in public expectations.
2. Conduct research and analysis using anticipatory crisis scenarios.
3. Determine relevant facts, research, and information using ethical tests.
4. Evaluate potential impact and set priorities using ethical analyses.
5. Develop strategy based on the ethical course of action.
6. Implement strategic communication.
7. Evaluate effectiveness.

The rest of this chapter explains each of these seven steps in detail as a way to manage pre-crisis situations with the utmost attention to ethics.

Although the implementation of ethical principles should pervade *each* step in the pre-crisis issues management process, ethical considerations were added in the steps that required the most intensive ethical analyses. Step number three of the process is entirely new—a specific step of determining relevant facts, research, and information as related to an issue (Stacks & Michaelson, 2010) is important to explicate separately because of the focus on ethics that it requires. Further, the order of steps in the process was reorganized to accurately indicate how modern organizations are conducting issues management, based upon the findings of several studies (Bowen, 2000, 2005; Bowen & Heath, 2005; Heath & Palenchar, 2009; Lauzen, 1997).

Monitoring or scanning the environment to identify issues remains the first step, but research and analysis take place shortly after the identification of an issue to help determine the extent of the issue, to find what information is needed, and to help assess the potential impact of the issue before priorities among issues can be set. Scenario building is also a part of research, in seeking information to anticipate and assess potential for crises as well as ways to lessen the likelihood of their occurrence. It is important to note that research is often an ongoing part of an issues management campaign and is normally revisited throughout the process rather than being confined to steps two and seven. For the sake of a parsimonious model, steps two and seven are used to indicate where public opinion research measurements are normally initiated, but other types of research are usually ongoing. For example, issue monitoring is in itself a form of research that is conducted throughout the process.

Involving Ethical Analyses

The challenges of identifying and defining issues for pre-crisis planning—anticipatory scenario building—involve many ethical concerns. Those ethical concerns are at the center of effective pre-crisis planning because they center on how an organization can build trust and maintain positive relationships with publics. Without trust, relationships can be permanently ended or inextricably damaged by a crisis. Engaging in ethical analyses along with crisis planning allows an organization to thoroughly consider both its *consequences on publics* and *responsibilities to publics*, thereby maintaining and enhancing trust.

The remainder of this chapter discusses each step of the process and will elucidate the type of moral philosophy that can be applied in that step in order to explicitly involve ethics in the decision-making process. Moral philosophy is a rigorous base upon which to build an ethical analysis because it offers a rationalistic approach, uses research in its decision-making, seeks to provide a consistent means of moral analysis, and offers a rigorous understanding of that which is deemed ethically good by most philosophers, societies, and publics. Moral philosophy offers a solid framework upon which the communicator can base decisions and offer a defensible explanation of those recommendations or decisions.

Monitoring the Environment for Issues and Trends Using Utilitarianism. In this first step of pre-crisis communication, the goal is to identify an emerging issue or problem long before it can become a crisis so that it can be resolved in a proactive and ethically responsible manner. Scanning the environment for emerging issues and trends refers to the information environment both inside and external to the organization. Scanning the environment for new issues and monitoring the development of identified issues can be accomplished in numerous ways. For example, an organization might subscribe to a service that routinely monitors media and website mentions of the organization. The organization might assign public relations professionals to routinely visit websites and blogs of activist groups and advocates, government organizations, regulatory bodies, industry associations, elected officials, and so on, looking for topics, concerns, problems, or trends that could eventually impact the organization. The organization will probably also conduct reviews of the transcripts from its toll-free telephone number and complaints or inquiries by tweets, telephone, postal mail, Facebook pages, and e-mail. Aggregating this information allows the issues manager a large resource of various types of communications from which to identify patterns that could indicate an emerging issue or a trend in the environment.

Once an issue is identified, it is monitored, and more research is normally conducted to help understand the extent of the problem and the potential for damage to the organization. Damages can be in the form of the regulatory environment, the reputation of the organization, changes to its business model, product recalls or other losses of brand equity, and the like. Proactively identifying issues is an essential component of pre-crisis communication because it gives the organization time to conduct research, engage in thoughtful analyses, and develop a strategic plan to keep the issue from becoming a crisis. In this manner, the process can save the organization a great deal of money and can sometimes even prevent the bankruptcy or potential collapse of an organization.

To help conduct issues scanning and monitoring in an ethical manner, utilitarian moral philosophy is of great value. Utilitarian philosophy is a consequentialist-based paradigm, meaning that the decisions made using it are based on the consequences, or potential outcomes, that result from an action. In this case, the action can be simply doing business within a certain industry, or it can be as complex as implementing new changes imposed by government regulators. The action is what the issues manager can use to analyze and determine the publics or stakeholders upon whom the action has consequences. One can create a consequence map, similar to a stakeholder map, that shows the publics most affected by the action the organization is undertaking (or considering undertaking). Philosopher Jeremy Bentham and his student John Stuart Mill created utilitarian philosophy, which held that happiness or the creation of happiness was the ultimate good. Mill (1861/1994) argued, "Actions are right in proportion as they tend to promote happiness, wrong as they tend to produce the reverse of happiness" (p. 201).

Delving further into utilitarian theory provides the issue manager a useful means of analysis for strategic issues management in pre-crisis situations. Although happiness is an abstract concept, utilitarian theory was refined to include the principle of maximizing positive outcomes while minimizing negative outcomes. In other words, the greatest good for the greatest number among publics, while reducing negative impact, is the ethical course of action that serves the public interest. Harm should be minimized, and decisions should promote the public interest. According to Elliott (2007), Mill advised that providing for our own happiness "does not violate what is due others" (p. 108). Ethicists in the utilitarian tradition encourage us to examine the potential impacts of our actions as a means for determining what is ethical.

Through examination of the consequences of organizational behavior on publics and weighing of the greatest good for the greatest number and minimizing any harm, we can use utilitarian theory to both scan for emerging issues and most carefully monitor those that have potential consequences on publics. Utilitarian theory can help segment publics into those that bear most consequences from organizational actions so that we will know who to prioritize and with whom to establish pre-crisis communication strategies. It can assist the issues manager in ascertaining what decisions have the most potential for benefiting the public interest and warn us of the consequences with the most potential for negative outcomes or harm.

Utilitarian ethics allows the issues manager to understand what publics and issues warrant the most communication and also warrant higher priority levels in the organization's management. The careful application of utilitarianism creates ethical pre-crisis communication strategy because it allows organizations to thoroughly consider the consequences that their activities have on their publics and to proactively act in a manner that seeks to minimize harm and maximize benefit.

Ethics of Conducting Research and Analyses: Categorical Imperative. In guiding the conduct of issues research, the ethical theory that issues managers should find most helpful is derived from Immanuel Kant's (1724–1804) philosophy. His moral philosophy is known as *deontology* because it is based on duty. In order to be ethical, this approach holds that a decision must be made entirely from the basis of rational analysis. This rational analysis must be free of personal bias, fear of retribution, and the like so that it is only influenced by the merits of moral principles underlying positions, arguments, and viewpoints. The issues manager is required to detach him or herself from personal connections to the situation and analyze it from a 360-degree objective approach involving all viewpoints—both internal to the organization and of external publics. In this exercise, every viewpoint is considered equally so that the underlying principle can emerge. Consequences are not entirely eschewed in the deontological approach; they are simply one of the many factors that must be considered along with the viewpoints of multiple publics and organizations. The most rationally compelling moral principles that emerge from the analysis are the ethical course of action to follow.

In using this deontological approach to guide research, the issues manager is free to explore all potentially worthy topics until the rational analysis is satisfied. The organization's viewpoint is not privileged over that of its publics, so the objective analysis can lead to a universal truth rather than one simply preferencing the desires of management. Universal moral principles are those that, in Kant's view, any rational decision-maker can understand in the value of the argument. One example is the universal principle of *honesty*; honesty is so entrenched across cultures and societies that we assume a communication is honest until we have a reason to suspect otherwise.

Basing research and analyses on a universal, rational vantage point allows the organization to thoroughly understand and consider all sides of an issue before a decision is made. That understanding can be enormously valuable in helping to formulate comprehensive organizational policies that help to prevent crises and maintain long-term relationships with publics. Those relationships can be exceedingly valuable to an organization in the event of a crisis. Further, conducting multi-perspective research allows publics to have some input into the information the organization has at hand for later use in strategy formation. The result should be more enduring and more comprehensive decisions that prevent crises and enhance relationships with publics.

Ethical Guidelines for Determining Facts: Law of Justice. In determining what facts are relevant to an issue, the ethical theory that issues managers should find most helpful is again derived from Immanuel Kant's deontological philosophy. The second form of Kant's imperative is especially germane to determining what facts and information publics need and should have in order to make informed judgments. This formula of Kant's (1785/1964) categorical imperative reads, "Act in such a way that you always treat humanity, whether in your own person or in the person of any other, never simply as a means, but always at the same time as an end" (p. 96). In essence, Kant's test requires that the dignity and respect owed toward publics is maintained by treating them not as a means to an organization's ends but as ends in themselves.

This formula of the categorical imperative is also known as Kant's *law of justice* (Sullivan, 1989). The test emphasizes the quality and moral worth of all publics, requiring the issues manager not to preference self-interest or organizational interest over the interests of publics. Kant's law of justice is particularly germane to public relations and issues management because it obligates the decision-maker to take a 360-degree view—to equally consider the interests of *all* publics involved in a decision (Bowen, 2004a). This test helps us, in the words of one philosopher (Baron, 1995), to "avoid the egocentric, self-satisfied approach" (p. 43) so that issues can be managed from the basis of moral duty to do what is right judged by analysis alone.

Although all points of view will likely not be determined to have equal worth in the decision, the test requires the issues manager to consider each of those points of view because of the dignity and respect morally owed to each potentially affected public. In this approach to ethical philosophy, it is the autonomous rationality, or the independent analysis and decision-making ability, that makes each person or public worthy of moral consideration. That rational autonomy obligates the issues manager to rationally consider the viewpoints of all publics. It also places all considerations on an equal footing where their merit can be determined by rationality alone rather than by what would serve our own self-interest or other forms of bias. Kant's test is a very powerful one for inclusion in issues management to help determine what facts are relevant to an emerging issue.

In modern-day terms, we might rephrase the categorical imperative formula to address the current business context of issues management and crisis management. Emphasizing the same moral principles in Kant's categorical imperative but using modern-day terminology, we would arrive at these tests:

- Does this decision maintain the dignity and respect of all involved publics?
- Do publics have access to all the information they need to make an informed decision?
- Are we, as an organization, preferencing our own self-interest by using our publics as a means to accomplish our goals? Or, do we respect the dignity of those publics and treat them as ends in themselves?
- Have we considered the divergent viewpoints of publics about this issue? What are their interests, values, and priorities?
- Is there a different interpretation of what facts might be relevant in this issue?
- In determining the relevant facts, have we considered all the possible alternate points of view? Have we truly taken a 360-degree view of the issue and its facts?
- Is the decision a rational one, meaning does it value the dignity of publics even if they do not agree with us, and can they see our point of view rationally even if they do not agree?

- Does this decision value long-term relationships with publics over short-term goals?

The analysis of the issue would not be limited to these questions but should also be open to input and feedback from various publics about their information needs, values, and priorities. By employing a practical form of Kant's categorical imperative, these questions provide a more inclusive model for determining the facts and information needs than traditional approaches to issues management in which organizational needs are the primary and most important consideration. As related to an issue, the questions provide an excellent start when used in conjunction with the other ethical theories reviewed in this chapter. These questions were also rephrased for a younger audience in a later study testing the results of using the model with millennials (Curtin, Gallicano, & Matthews, 2011).

Maintaining the dignity and respect of publics is essential for an organization that wishes to create long-term success. If publics can see that an organization considers their points of view and allows them the moral autonomy to make their own decisions based on the facts relevant to an issue, they will ultimately become engaged in long-term relationship building with that organization. Even publics that do not support the mission or vision of the organization are less likely to try to hinder it through their own public policy initiatives when they know that their point of view has been included in the decision-making process.

Evaluate Potential Impact; Set Priorities via Rule Utilitarianism. In Crable and Vibbert's (1985) view, issues enter the public policy arena when organizations fail to solve problems recognized as significant by publics. That very definition implies that our pre-crisis planning has failed to access the potential for damaging issues to impact both the organization and the publics who care about the issue. To make pre-crisis planning or issues management the most effective it can be, potential impacts and priorities must be accurately assessed or—as a minimum—we can seek to reduce uncertainty in our decision-making process. Potential issues are always fraught with uncertainty that can be reduced by thoroughly examining the consequences of potential decisions with regard to their impact on publics.

The long-term relationship building inherent in the communication discipline is not always easily understood by management. A more short-sighted manager may ask, "Why should the organization make decisions based upon what publics want?" However, shared control of issues and mutual benefit from solutions, whenever possible, leads to relationships in which both parties are satisfied, rather than consistently seeking redress. Grunig and Repper (1992) explained the idea of consequences on publics by writing, "…public relations makes organizations more effective by developing relations with stakeholders and the internal or external environment that constrain or enhance the ability of an organization to accomplish its mission" (p. 118). When issues managers carefully consider the consequences of their pre-crisis decisions, the organization can expect less opposition from publics and, sometimes, even an enhanced ability to move toward its goals.

Utilitarianism is a marvelous fit for a moral analysis that would examine the implications of issues and their potential solutions on publics. Utilitarianism, as discussed above, is a consequentialist theory, meaning that the determination of what is ethical is based upon the potential outcomes of the decision. Those outcomes can be examined by using an act utilitarian framework.

Potential outcomes of emerging issues can be predicted by using a *rule utilitarian approach*, and issue priorities can be set according to the ethical obligation of managing the greater good for the greater number while minimizing the harm caused by an issue. A rule utilitarian approach is one in which the utilitarian framework applied seeks to group actions by classes of like behaviors and like outcomes. Utilitarianism holds that we must consider all persons potentially affected by a decision, but in rule utilitarianism the decision is actually a class of action involving general outcomes rather than specific consequences. Using a rule utilitarianism approach, as opposed to an act utilitarianism approach involving specifics, allows the issues manager to assess the potential consequences of a decision in an expeditious manner. DeGeorge (1999) explained: "Why favor the rule utilitarian rather than the act utilitarian approach? The answer is that we cannot know all the consequences of a particular act, nor can we know in advance, and with certainty, many of the specific consequences of such an act" (p. 62). In other words, a rule utilitarian approach minimizes the guesswork that an issues manager must make by employing now common outcomes and like cases to evaluate potential consequences.

Rule utilitarianism allows a type of action to be judged in accordance to the amount of good versus harm it generally produces. For example, it is common for governments to employ a rule utilitarian approach to lawmaking because it benefits the majority of the society and creates more positive than negative outcomes. DeGeorge (1999) pointed out that it is a rule utilitarian analysis that results in most penal codes, attaching negative consequences or sanctions to a class of actions in the form of sanctions. Sanctions are attached as negative consequences for the perpetrator so that the positive outcomes do not provide more benefit than negative outcomes. In the organizational sense, a pre-crisis rule utilitarian analysis of potential consequences can provide certain classes of actions that are deemed to create benefits and others that are deemed unwise because they produce a mixed ratio of benefit to harm or of negative consequences. Harm must always be justified by promoting the aggregate good of the overall community (Elliott, 2007).

In seeking to determine potential outcomes, pre-crisis situations would be judged according to the rule-based utilitarian principles of justice that Mill (1861/1957) espoused. Elliott (2007) provided a simple synopsis of these principles: It is unjust to deprive people of their legal rights; moral rights and civil rights should be maintained; it is just to receive the good one deserves or the evil one deserves (justifying the punishment of criminals); one should not break faith; and one should treat matters objectively or impartially (pp. 108–109). Despite their treatment as principles, Mill held that they were not absolute rules but guidelines that the decision maker should consider before entering the utilitarian calculus of weighing benefits versus harms.

Potential issues that an organization might face can be analyzed according to their adherence to utilitarian principles, meaning the outcomes of similar cases in the past—the moral rules involved in a class of actions such as honesty, obligation, fairness, or even fair competition in the marketplace. *The utilitarian calculus simply means that the amount of good or positive outcome predicted by a consequence indicates whether that action is ethical.* Choices among actions necessitate that the decision with the most positive outcomes and least negative outcomes is the ethical choice.

Rule utilitarianism also holds that an organization should rank the issues—in terms of priority and risk—that have the potential to create the most negative consequences (for a complete discussion of the ethics of risk management, see Bowen, 2009a). Potential issues can be ranked according to their severity by the amount of harm they cause. Crisis situations are defined by the amount of negative consequences produced, and it is logical to seek to reduce negative consequences as an ethical outcome of pre-crisis management. *Creating the greatest amount of good can often be defined in a crisis situation as the decision that minimizes negative or bad consequences.*

A clear assessment of the consequences of pre-crisis issues will allow the decision maker to thoroughly understand the stakeholders and publics potentially impacted by various responses to the issue as well as to find which response serves the greater good. Creating the greatest good for the greatest number, while seeking to minimize harms, can help the organization to respond to pre-crisis issues in an ethical way that has thoroughly considered the impacts its actions create. Further, that consideration can serve as the basis for helping to build long-term, satisfied relationships between organizations and publics. Once potential consequences are evaluated and priorities are set, issues managers use research (e.g., Stacks, 2011; Michaelson & Stacks, 2011) to help formulate strategy to avert crises.

Developing Strategy Based on Act Utilitarianism. Decision options for resolving pre-crisis issues are normally discussed by the issues manager in conjunction with the dominant coalition or leaders of the functional areas related to the problematic topic. Various options for strategy will be weighed according to their consequences based upon research on the topic. That research is both specific to the actual issue inside the organization (e.g., product related, scientific, brand specific, and so on) and public opinion research that is conducted by the public relations department in order to assess knowledge and attitudes about the issue among publics. Buchholtz, Evans, and Wagley (1994) argued, "To a large extent the success of the company will depend on how well it formulates its strategy in light of an evolving environment, how well it defines an articulated strategy, and how well it ensures implementation of the strategy" (p. 21). Informing pre-crisis planning and decision making with as much research as possible is a wise move because it allows the ultimate strategy to be formulated with as much information on accurate consequences as possible.

In developing the organization's strategy, priorities should have emerged in the prior step based on potential impact of the issue. Of those issues deemed important enough to warrant a strategic response, issue option strategies should be developed based upon practicable responses and mission, ethics, scenario building, input from publics, and resources (i.e., cost) (Buchholz et al., 1994). The other considerations for strategy formation in pre-crisis management are discussed throughout this volume, so the ethics of strategy formation will be focused upon here. As Stacks (1995) explained, "To engage the public relations strategic plan, that plan must identify and prioritize the audiences within the various publics" (p. 26).

Act utilitarianism is an ideal match for the forecasting and scenario building required in managing a pre-crisis strategy. The issue options at hand, or the potential strategies for resolving the pre-crisis issue, can be considered individually, in all of their specificity, with regard to their potential consequences. Act utilitarianism goes a step further than does rule utilitarianism because it requires the decision-maker to consider the specific consequences of the actual decision at hand in all of its intricacies. The utilitarian calculus is still the guide for ethical behavior, meaning that decision should be based on that which maximizes the greatest good and minimizes harm or negative outcomes. However, each decision option is weighed separately to determine the potential consequences of implementing the action.

Act utilitarianism is helpful in determining how publics will respond to an organization's activities. It allows the issues manager to look toward future outcomes and implement the scenario that benefits the greatest number of people and harms the fewest. Although this approach is difficult to implement because outcomes are often unpredictable, a sizable amount of reliable research based on a variety of data gathering methodologies that triangulate data on the issue (Tashakkori & Teddlie, 1998) can help to offset the uncertainty prevalent in act utilitarianism. The decision-maker should make certain not to neglect or harm the rights of others (legal rights or civil rights) by an action under consideration, remembering that any harm must be warranted and deserved according to Mill's principles of justice (1861/1957). Further, the specificity of act utilitarianism allows organizations to effectively know and understand the consequences of their actions before those actions are taken, charting a more responsible course for the greatest number of people.

Implement Strategic Communication. Effective pre-crisis planning requires that an organization implement and test a strategic issues management plan in order to solve issues (potential crises) before they negatively impact the organization or its publics. Following the above step, developing strategy based on an analysis via act utilitarianism, an organization then implements the chosen strategy through a strategic communication campaign, sometimes also called an issue advocacy campaign (Heath, 1997)—specifically where matters of public policy and legislation are involved.

Despite the in-depth analyses in the prior steps, an issues manager must also answer the question: "How should the strategy be communicated in order to be ethical?" In the ethical analysis of the implementation of a strategic communication campaign, it is vital that an interest in evaluating the consequences on public policy be maintained by the issues manager. In other words, how can the correct strategy be implemented while maintaining the deontological rationality required, dignity and respect due to publics, a rule utilitarian principle of good, and then act utilitarian conviction that the strategy produces the greatest good for the greatest number?

The answer to that complex question lies once again in deontology, or duty-based and rationalistic moral principle. The implementation of any issue advocacy campaign or strategic communication campaign intended to address pre-crisis concerns should be based on a moral conviction of pure intention, or what Immanuel Kant called the "morally good will." The good will, in Kant's philosophy, required that not only are rationality and the dignity and respect of publics maintained but also that any decision proceeds from a basis of obligation to do the right thing alone (Sullivan, 1989, 1994).

Kant maintained that the only incorruptible human virtue was pure intention or the good will (Sullivan, 1989). The good will is also known as the third formula of Kant's categorical imperative, obligating all decision-makers categorically, rather than conditionally, to act from good will because they understand, through a rational analysis, their moral duties—what a later moral philosopher called a coherent theory of goodness (Paton, 1927). All people are obligated to act morally because they have the higher reasoning ability that comes through rationality. Kant (1785/1993) explained his conception of pure intention: "Nothing can possibly be conceived in the world, or even out of it, which can be called good without qualification, except a GOOD WILL" (emphasis in original, p. 154).

A good intention should drive the strategies of an organization in pre-crisis management and in the issue advocacy campaigns that seek to prevent crises. Otherwise, such communications should not be implemented. The organization cannot be bent upon protecting its own interests alone; in fact, such an intention would make the action patently *unethical*. Because a good will is the only thing that is incorruptible, it is the only ethical basis from which one can act and have a decision still deemed ethical from multiple perspectives. Therefore, an organization must implement its strategic campaigns and communication of those campaigns from an intention of simply doing the right thing. That means that an organization must (1) want to actively prevent the crisis and (2) resolve any dilemmas or issues because a moral principle based on rational analysis indicates that anyone in a similar situation would conclude that way to resolve the issue is the right thing to do. Acting from the intention of doing something because it is right is morally praiseworthy; acting from self-interest is not. Organizations must prevent crises *because they are good will agents within society* who are responsible to their publics, governments, and other interested stakeholders (Bowen, 2009a), acting from a rationale that it is the right thing to do. Kant's third form of the categorical imperative specifying that we should act from good will alone is, indeed, a difficult test to pass. However, the analytical and well-intentioned organization, which pays attention to ethical analyses, should be able to act based on those analyses from utilitarian and deontological perspectives—from an interest of doing the right thing.

Evaluate Effectiveness: Are the Interests of Publics Maintained? Evaluating the effectiveness of pre-crisis activities means not only measuring if a crisis was averted but also measuring whether affected publics are satisfied and understand the perspectives of the organization. Evaluation research must be conducted not only to measure the efficacy of the strategic communication and the reduction in crisis likelihood but also the ethical implementation of the strategy. Evaluation research can help to refine the organization's strategy, emphasizing its strengths and compensating for or eliminating weaknesses. Even after the ethical examinations provided by deontological and utilitarian philosophy, the intentions, responses, and consequences of decisions are unpredictable. Therefore, measuring these outcomes allows an organization to understand them and respond immediately rather than waiting until damage may have been done.

Implementing strategy ethically not only helps to protect the organization's reputation from harm but also helps to insulate the revenue stream and the business model from potential crises while maintaining long-term relationships with publics. In seeking to maintain relationships, the organization has a moral intention that is based on good will and that also serves its longer-term interests of staying effective and achieving its organizational goals. In reviewing the effectiveness of a pre-crisis issue campaign, an organization should re-evaluate both its moral intention of good will and effectiveness in achieving goals, rational judgments, and the predicted consequences of decisions. Were the consequences for publics different than expected, or were reactions from publics unanticipated? Evaluation allows an organization to respond to unexpected outcomes with more predictable consequences.

Moral Education

To implement these standards from utilitarianism and deontology, public relations professionals need to be trained in ethics and moral decision-making. At least a basic understanding of both utilitarian philosophy and deontological philosophy is necessary to conduct a rigorous and thoughtful methodical analysis of a pre-crisis situation. Such situations could unfold into issues that damage the organization's reputation and cause financial harm, sometimes to the point of bankruptcy or dissolution. It is worth companies investing in *required* ethics and moral philosophy training at the university level and the ongoing professional level to help organizations prevent crises resulting from moral and ethical lapses and their negative impacts on society and publics within that society.

Issues managers are being asked more commonly than ever before to act as ethical counsels within their organizations (Bowen, 2008). In fact, research indicates that ethical dilemmas also offer the occasion to gain membership in the dominant coalition or top decision-making core of an organization for strategic public relations and issues management (Bowen, 2009b). Ethics training is needed in order to provide that type of counsel to the dominant coalition (Bowen et al., 2006). Training guides and books such as Bowen, Rawlins, & Martin (2010) offer additional insight into the analysis and resolution of ethical dilemmas. That sort of knowledge can be of great assistance to issues managers in seeking to prevent crises.

Although it may seem an unusual combination to use both utilitarian theory and deontological philosophy to resolve moral dilemmas—and the combination is an admittedly unique perspective—they provide a complex and sound analysis for issues managers. Even Kant acknowledged the value of consequences as one element in a rational analysis when he said, “There can be no will without an end in view” (Sullivan, 1989, p. 25). By combining deontological and utilitarian frameworks, the rational analyses that result can be based on both moral principles and consideration of the consequences that an organization has for its publics. Consideration of those consequences should allow that organization to build and maintain more satisfactory relationships with publics over the long term. It should also allow the organization to engage in more accurate scenario building and understand its environment thoroughly before acting.

Conclusion

The thorough consideration of the categorical imperative and deontological moral principles allows organizations to satisfy their obligations as socially responsible entities within society, contributing not only to their own success but to the equality, dignity, and respect of those within that society—be they publics, other organizations, the self, or government. Finally, by engaging in ethical pre-crisis management and preventing crises, the organization itself becomes an ethically responsible member of society, contributing to the social good by fulfilling its duty to become the most excellent organization it can be.

References

Avery, E. J., & Kim, S. (2009). Anticipating or precipitating crisis? Health agencies may not be heeding best practices advice in avian flu press releases. *Journal of Public Relations Research, 21*(2), 187–197.

Baron, M. W. (1995). *Kantian ethics almost without apology*. Ithaca, NY: Cornell University Press.

Bowen, S. A. (2000). *A theory of ethical issues management: Contributions of Kantian deontology to public relations' ethics and decision making.* (Unpublished doctoral dissertation). University of Maryland, College Park, MD.

———. (2004a). Expansion of ethics as the tenth generic principle of public relations excellence: A Kantian theory and model for managing ethical issues. *Journal of Public Relations Research, 16*(1), 65–92.

———. (2004b). Organizational factors encouraging ethical decision making: An exploration into the case of an exemplar. *Journal of Business Ethics, 52*(4), 311–324.

———. (2005). A practical model for ethical decision making in issues management and public relations. *Journal of Public Relations Research, 17*(3), 191–216.

———. (2008). A state of neglect: Public relations as 'corporate conscience' or ethics counsel. *Journal of Public Relations Research, 20*(3), 271–296.

———. (2009a). Ethical responsibility and guidelines for managing issues of risk and risk communication. In R. L. Heath & H. D. O'Hair (Eds.), *Handbook of risk and crisis communication* (pp. 343–363). New York: Routledge.

———. (2009b). What communication professionals tell us regarding dominant coalition access and gaining membership. *Journal of Applied Communication Research, 37*(4), 418–443.

Bowen, S. A. & Heath, R. L. (2005). Issues management, systems, and rhetoric: Exploring the distinction between ethical and legal guidelines at Enron. *Journal of Public Affairs, 5*, 84–98.

Bowen, S. A., Heath, R. L., Lee, J., Painter, G., Agraz, F. J., McKie, D., & Toledano, M. (2006). *The business of truth: A guide to ethical communication.* San Francisco, CA: International Association of Business Communicators.

Bowen, S. A., Rawlins, B., & Martin, T. (2010). *An overview of the public relations function.* New York, NY: Business Expert Press.

Buchholz, R. A., Evans, W. D., & Wagley, R. A. (1994). *Management responses to public issues: Concepts and cases in strategy formulation* (3rd ed.). Upper Saddle River, NJ: Prentice Hall.

Chase, W. H. (1984). *Issue management: Origins of the future*. Stamford, CT: Issue Action Publications.

Coates, J. F., Coates, V. T., Jarratt, J., & Heinz, L. (1986). *Issues management: How you can plan, organize and manage for the future*. Mt. Airy, MD: Lomond.

Coombs, W. T. (1999). *Ongoing crisis communication: Planning, managing, and responding*. Thousand Oaks, CA: Sage.

———. (2007). *Ongoing crisis communication: Planning, managing, and responding* (2nd ed.). Thousand Oaks, CA: Sage.

Crable, R. E., & Vibbert, S. L. (1985). Managing issues and influencing public policy. *Public Relations Review, 11*(2), 3–16.

Curtin, P. A., Gallicano, T. & Matthews, K. (2011). Millennials' approaches to ethical decision making: A survey of young public relations agency employees. *Public Relations Journal, 5*(2), 1–22.

DeGeorge, R. T. (1999). *Business ethics* (5th ed.). Upper Saddle River, NJ: Prentice Hall.

Elliott, D. (2007). Getting Mill right. *Journal of Mass Media Ethics, 22*(2&3), 100–112.

Gaunt, P., & Ollenburger, J. (1995). Issues management revisited: A tool that deserves another look. *Public Relations Review, 21*(3), 199–210.

Goldberg, M. (Ed.). (1993). *Against the grain: New approaches to professional ethics*. Valley Forge, PA: Trinity Press International.

Grunig, J. E., & Repper, F. C. (1992). Strategic management, publics, and issues. In J. E. Grunig (Ed.), *Excellence in public relations and communication management* (pp. 117–157). Mahwah, NJ: Lawrence Erlbaum Associates.

Heath, R. L. (1997). *Strategic issues management: Organizations and public policy challenges*. Thousand Oaks, CA: Sage.

Heath, R. L., Lee, J., & Ni, L. (2009). Crisis and risk approaches to emergency management planning and communication: The role of similarity and sensitivity. *Journal of Public Relations Research, 21*(2), 123–141.

Heath, R. L., & Nelson, R. A. (1986). *Issues management: Corporate public policymaking in an information society*. Beverly Hills, CA: Sage.

Heath, R. L., & Palenchar, M. J. (2009). *Strategic issues management: Organizations and public policy challenges* (2nd ed.). Thousand Oaks, CA: Sage.

Huang, Y. H. (2001). OPRA: A cross-cultural, multiple-item scale for measuring organization-public relationships. *Journal of Public Relations Research, 13*(1), 61–90.

———. (2008). Trust in relational commitment and corporate crises: The effects of crisis communicative strategy and form of crisis response. *Journal of Public Relations Research, 20*, 297–327.

Jaques, T. (2007). Issues management and crisis management: An integrated, nonlinear, relational construct. *Public Relations Review, 33*(2), 147–157.

———. (2009). Issue and crisis management: Quicksand in the definitional landscape. *Public Relations Review, 35*, 280–286.

Jin, Y., & Cameron, G. (2007). The effects of threat type and duration of public relations practitioner's cognitive, affective, and commented responses in crisis situations. *Journal of Public Relations Research, 19*(3), 255–281.

Kant, I. (1785/1964). *Groundwork of the metaphysics of morals* (H. J. Paton, Trans.). New York: Harper & Row. (Original publication 1785).

———. (1785/1993). Metaphysical foundations of morals (C. J. Friedrich, Trans.). In C. J. Friedrich (Ed.), *The philosophy of Kant: Immanuel Kant's moral and political writings* (pp. 154–229). New York: The Modern Library. (Original publication 1785).

Lauzen, M. M. (1997). Understanding the relations between public relations and issues management. *Journal of Public Relations Research, 9*(1), 65–82.

Lauzen, M. M., & Dozier, D. M. (1994). Issues management mediation of linkages between environmental complexity and management of the public relations function. *Journal of Public Relations Research, 6*(3), 163–184.

Michaelson, D., & Stacks, D. W. (2011). Standardization in public relations measurement and evaluation. *Public Relations Journal, 5*(2), 1–22.

Mill, J. S. (1861/1957). *Utilitarianism*. New York: The Liberal Arts Press.

———. (1861/1994). Higher and lower pleasures. In P. Singer (Ed.), *Ethics* (pp. 201–205). Oxford: Oxford University Press.

Mitroff, I. (2003). *Crisis leadership: Planning for the unthinkable*. New York, NY: Wiley.

Murphy, P. (1996). Chaos theory as a model for managing issues and crises. *Public Relations Review, 22*(2), 95–114.

Olaniran, B. A., & Williams, D. E. (2001). Anticipatory model of crisis management: A vigilant response to technological crises. In R. L. Heath (Ed.), *Handbook of public relations* (pp. 487–500). Thousand Oaks, CA: Sage.

———. (2008). Applying anticipatory and relational perspectives to the Nigerian delta region oil crisis. *Public Relations Review, 34*, 57–59.

Paton, H. J. (1927). *The good will: A study in the coherence theory of goodness*. New York: Macmillan.

Renfro, W. L. (1993). *Issues management in strategic planning*. Westport, CT: Quorum.

Sims, R. R. (1994). *Ethics and organizational decision making: A call for renewal*. Westport, CT: Quorum.

Stacks, D. W. (1995). Travel public relations models—public relations and travel: A program approach to communication in the 1990s. *SPRJ, 1*(2, Winter).

———. (2011). *Primer of public relations research* (2nd ed). New York, NY: Guilford.

Stacks, D. W. & Michaelson, D. (2010). *A practitioner's guide to public relations research, measurement, and evaluation.* New York, NY: Business Expert Press.

Sullivan, R. J. (1989). *Immanuel Kant's moral theory*. Cambridge: Cambridge University.

———. (1994). *An introduction to Kant's ethics*. New York: Cambridge University.

Tashakkori, A., & Teddlie, C. (1998). *Mixed methodology: Combining qualitative and quantitative approaches*. Thousand Oaks, CA: Sage.

Chapter 5

Recognizing, Anticipating, and Preventing Ethical Misconduct Disasters

Robert C. Chandler

The author introduces us to the concepts of strategic ethical integrity and integrity management. This chapter presents essential work designed to help large organizations avoid the pitfalls of ethical misconduct disasters, which can be extremely damaging in terms of reputation, market share, and profit. This chapter takes us through the process of identifying signs of potential danger with ethical crises, scandals, and other forms of ethical misconduct. The author then offers a comprehensive overview of informal and formal means for employing integrity management practices designed to guide organizational members toward legally sound practices that adhere to corporate and professional codes of conduct.

Introduction

This chapter reviews the concepts of strategic ethical integrity in addition to the measures required for recognizing the warning signs of ethical crises, scandals, and ethical misconduct disasters. The techniques and tactics for anticipating ethical problems are reviewed, including compliance programs and ethics training initiatives for employee behaviors and decision-making. The process and benefits of conducting an Ethical Conduct Audit© (ECA) are also discussed. Finally, guidelines for preventing ethical crises and ethical misconduct disasters are presented. These include a forward-looking perspective for sustaining integrity continuity in an organization as well as key steps to reach sustained organizational integrity continuity.

Overview

Scandals, accidents, product problems, criminal activity, deception or fraud, misconduct, harassment, discrimination, financial or regulatory improprieties, malfeasance, misappropriations, or ethical breaches cannot only damage the reputation of corporate executives but can wreak financial havoc on the value of a company's brand assets. The corporate ethics scandals of the past decade have demonstrated that misconduct can disrupt or destroy companies. The increased regulatory scrutiny and potential punitive sanctions for misconduct have raised the stakes for senior management, corporate executives, and boards of directors (LeClair, Ferrell, & Fraedrich, 1998). It is time to come to terms with the stark reality that corporate integrity and the ethical resiliency of a company must be considered aspects of strategic continuity planning.

There appears to be a need for a concern about ethical dispositions in both society and, more specifically, in the workforce. The results of a national study of 12,000 high school students included the following surprising findings: 74 percent admitted cheating on an exam at least once in the past year; 38 percent admitted shoplifting at least once in the past year; and 37 percent admitted that they would lie "in order to get a good job" (Josephson, 2002).

In addition, the "inherent ethics" of the "good moral people" that a company hires include the 76 percent of Master of Business Administration (MBA) graduates who reported that they were willing to commit fraud to enhance profit reports to management, investors, and the public (Lazere, 1997). Fewer than 50 percent of employees believe their employers have high ethical integrity (Walker Information and Hudson Institute, 1999). Thirty percent of all employees currently report that they "know or suspect ethical violations such as falsifying records, unfair treatment of employees, and lying to top management" (Walker, 1999). Specifically, 41 percent of employees in the private sector and 57 percent of employees in the public/government sector are aware of ethical misconduct or illegal activities (Integrity and Ethics, 1998). Furthermore, 60 percent of employees state that they know, but have not reported, instances of misconduct in their organizations (Taub, 2002). Most employees cite these reasons for not coming forward about ethical misconduct in companies: the lack of confidentiality policies, fear that existing policies won't protect them, and the fear of "whistle-blower retaliation" (Taub, 2002).

Ethical behavior is a growing concern across society in general. It is no surprise that these issues are encroaching into the workplace on many different fronts. The times have changed since the days when one could uncritically assume that all employees are hired with a fundamental and rigid commitment to recognizing, understanding, and acting ethically in every possible situation. Furthermore, even good moral individuals may be influenced by reward systems, unique temptations, or unseen pressures that will affect their ethical decision-making in some situations.

Clearly, issues with integrity occur at a high rate. As a result, it becomes necessary for policies to be instated to set guidelines for how to handle these situations. Unfortunately, integrity continuity considerations are oftentimes ignored by senior management, specifically at the chief executive level. Research has revealed that 60 percent of chief executives and boards of directors did not engage in integrity continuity planning discussions or include such considerations in strategic planning. Additionally, 57 percent of companies have never incorporated integrity continuity planning at the strategic executive or board level. More than half of all businesses fail to assess ethical misconduct risks and plans to ensure integrity continuity. An example of this is a statistic that states that 54 percent of companies do not have employee ethics compliance measurement in their performance appraisal criteria.

Integrity Management

When devising strategy concerning business policies and organizational structure, a multitude of factors are analyzed in terms of their importance. Prioritizing planning components allows businesses to understand where their needs lie and what areas of organization must be reformed. An essential factor to any business continuity plan is ethical integrity. Creating and sustaining organizational ethical integrity are critical for avoiding, mitigating, and surviving organizational scandals and unethical disasters. Ethical issues must be on the organizational agenda. Because ethics should play a significant role in all aspects of business, planning must extend beyond issues with compliance and discipline in order to manage integrity. Ethical management is necessary not only because it is a matter of legal jurisdiction but also—perhaps more importantly—because it is morally right. Businesses that make ethics a priority raise awareness of integrity issues for employees, in turn causing them to create personal ethical standards to abide by. Those companies without ethics policies fuel an atmosphere where employees perceive that their work is separate from ethical guidelines and, as a result, exploit other value systems that encourage them to deviate from integrity-based behavior.

Various components are essential in maintaining the effectiveness of integrity management within business organizations. Formal and informal ethical systems should be synthesized in order to meet legal criteria while simultaneously reaching ethical standards set by the company. In order to ensure that these policies are enacted successfully, a rewards system should be put into place to encourage employees to preserve ethical behavior in the workplace. Additionally, these guidelines should maintain simplicity and coincide with the collective moral framework of the employees. These policies should provide a means for consistent honest decision-making by employees at all levels. These decisions should be fair and appropriate. Organizational integrity ranks high in importance because of its ability to diminish the effects of unethical scandals. Ethical misconduct is damaging to the credibility of any organization and should be avoided in order to maintain the positive image that a company strives hard to create. Failures in integrity policy are easy for outsiders to observe; as a result, news of these breakdowns spreads quickly and successfully thwarts any prospects the affected company may be pursuing. Just as natural disasters can depreciate and destroy the value of a company, so can ethical violations. Brand names and company images can be forever ruined because of integrity mishaps. Preventing this type of breakdown is essential in preserving the appearance of any business organization. It is clear, then, that ethical policies must be implemented to prevent outsider attacks on integrity in the future.

Organizational integrity is the first line and fundamental strategic planning area to minimize the risks of major unethical scandals. Ethical misconduct disasters and scandals are serious, costly risks to the continuity and survival of a business. Evidence can be found almost daily in the news headlines that breakdowns of integrity collectively cost businesses billions of dollars in litigation, fraudulent financial acts, increased costs, fines, reputation and image damage, customer/client trust, lost sales, and recovery costs.

Strategic Integrity Continuity

Most organizations have long acknowledged that business continuity planning is an essential priority in order to effectively anticipate, prevent, mitigate, and survive natural disasters, data loss, accidents, and deliberate malevolent acts. What many are only now discovering is that integrity continuity planning is also a duly diligent policy and business priority. Ethical issues must be on the strategic agenda. Such planning must go beyond compliance issues and reactive disciplinary policies to actually manage integrity. Integrity management should be a priority because it is legally required and is the right thing to do. Employees who know that certain workplace decisions, behaviors, and processes exist in an ethically judged context are more aware and motivated to act ethically. Employees who perceive such activities as existing detached from an ethical context or who utilize an alternative (unethical) value paradigm (i.e., financial or perceived performance) are less aware of ethical implications and more motivated to act unethically.

The more employees recognize when a decision or behavior involves questions of ethics, the more their consciences are raised about the importance of acting ethically and the more likely they are to behave in appropriate ways. Senior management must set the example and communicate the importance of professional integrity in the workplace in order to create a climate of integrity.

Integrity management includes creating formal and informal systems, which ensure that employees will act in ways that ensure legal compliance as well as enact the corporate code of conduct goals in ways that are consistently professional, ethical, and desirable. Integrity management is intertwined with managing the larger corporate culture and informal reward/motivation processes that impact employee decisions and behaviors in ways that transcend policies printed in a written code of conduct. In many instances, major ethical scandals have occurred, even in companies that have clear and explicit policies and codes of conduct. It requires more than merely having policies in place to avoid, mitigate, and survive these risks. Obviously, every business needs to strategically create structures and processes that fundamentally require all employees and managers to obey all legal requirements and regulations. Furthermore, common ethical and professional standards would obviously include assumptions that decisions and behaviors are conducted honestly and that employees and managers would never knowingly harm or do damage to fellow employees, stakeholders, customers, clients, or vendors by deception, misrepresentation, fraudulent report, coercion, conflict of interest, or other mordacious acts. Organizational integrity must be strategically managed, as are other mission critical strategic factors connected with business success and effectiveness.

Ethical Misconduct Disasters (EMD)

Ethical misconduct disasters are specific, unexpected, and non-routine unethical events or series of unethical events that create significant operational disruptions and threaten—or are perceived to threaten—an organization's continuity of operations (Brewer, Chandler, & Ferrell, 2006). While acknowledging that there is no universally accepted definition of an ethical misconduct disaster, it is arguable that such a crisis can negatively affect a company, a major business unit, the reputation and image of a brand, and the stakeholders invested in the organization. Financial performance can be damaged, and employees can be impacted negatively, resulting in litigation and regulatory responses. Media explosions result from this type of problem, serving to reinforce the negative images an organization is associated with. Instances like these can destroy an organization and threaten the very existence of the corporation itself (e.g., Enron).

An ethical misconduct disaster can also severely damage an organization's financial performance, impact employees, result in litigation and regulatory responses, and create a media circus that can destroy the public's basic trust or belief in an organization, its reputation, and its image. The erosion of public trust has been furthered by the seemingly endless series of scandals and ethical disasters.

Recent revelations about ethical misconduct scandals make the prospects of the "unthinkable ethical disaster" a realistic concern. Prudent strategic continuity planning must take into account the risks of ethical misconduct disasters along with other serious threats to business continuity. Recent headlines document the need for strategic integrity continuity planning. What comes to mind when the following corporate names are mentioned: Martha Stewart, Quest, Merrill Lynch, Tyco, Enron, WorldCom, Andersen, Sears, Mitsubishi Motors, United Way of America, Global Crossing, Adelphia, Smith Barney, Citigroup, or the Roman Catholic Church? Senior executives are facing prison sentences as a result of some of these scandals. In a Sarbanes-Oxley world, what prudent executive would ignore the risks of ethical scandals? Ethical challenges are on the rise and becoming more difficult to anticipate and successfully manage in the increasingly complex work environment. Although predicting ethical scandals in American business is not an exact science, one CFO.com projection forecasts up to 20 "major" business ethical misconduct disasters every year (Lazere, 1997).

Not every instance of an unethical decision that occurs necessarily explodes into a catastrophe for the organization. More frequently there are unethical acts, deception, bias, and questionable conduct occurring in most organizations. Sometimes these are relatively minor infractions whose consequences are limited to very localized areas. Other times, substantial breaches of ethics occur, but in many cases they are never exposed or questioned. Or, the checks and balances that exist can help resolve and regulate these to limit the prospect of major scandals and ethical misconduct disasters. In fact, businesses that effectively manage integrity can systemically absorb, react, and appropriately adjust to most breakdowns in conduct or decisions. The key is whether the organization has adequately planned to mitigate through prompt response, disciplinary action, appropriate disclosure, communication to the workforce, and public crisis management communication in order to manage these events so that they do not escalate into catastrophes. Misconduct, because of its severity, persistence, lack of quick and appropriate response, public scrutiny of the organization's mishandling of the event, or necessary involvement of legal or regulatory structures may escalate to a level where it is classified as a disaster for a business. Such an ethical misconduct disaster goes beyond the mere disruption of routine operations. It poses disastrous results that have potentially significant economic consequences or in some cases present a threat to the survival of the business.

Ethical misconduct scandals can arise from a variety of segments or levels of a company's operations. The major categories of such disasters generally include instances of harassment or discrimination, criminal or illegal activities, financial improprieties, customer deception, bribery or improper influence, regulatory violations, corruption, or undisclosed conflict of interest. These scandals can disrupt the operations of a corporation, damage performance, create entangling litigation, and destroy reputation or brand image. Creating guidelines to prevent these types of circumstances holds employees accountable to certain standards. Corporations must be increasingly held accountable for their efforts to educate, train, and enable employees to act ethically, legally, and with integrity in the performance of their duties.

Recognizing the Risk for Ethical Misconduct Disasters. Ethical scandals can be disruptive and destructive to a business, agency, or organization. Ethical breakdowns can disrupt operations just as significantly as natural disasters, such as Hurricane Katrina, potential data loss such as the Y2K threat, industrial accidents, and deliberate malevolent acts such as the 9/11 terrorist attacks or the Virginia Tech shooting. Such scandals can have enormous costs, impact a business, wound a reputation and brand, anger stakeholders, and potentially land senior management in prison. No company is immune from these threats. Prudent businesses must plan to manage integrity continuity by assessing their vulnerability to ethical disasters, taking proactive measures, and preparing their organizations to mitigate and survive when such scandals occur. Achieving consistent integrity is much more than simply ensuring legal and regulatory compliance. Proactive integrity continuity requires that a company have a fundamental strategic commitment to integrity, be aware of intrinsic risks, and build a culture of integrity that ensures that company employees are likely to make ethical decisions and employ ethical behavior. Effectively managing integrity is dependent on creating an ethical context in which employee decisions and behaviors are aligned with the organizational mission as well as critical goals of other professional, ethical, and strategic priorities. Managing integrity (as in managing people) involves managing employees' perceptions of situations and shaping an understanding of how they should act, behave, and decide in the various situations in which they must perform their work.

Ethical misconduct scandals can spring from any segment or level of a company's operations. As mentioned earlier, the major categories of such disasters typically include instances of harassment or discrimination, criminal or illegal activities, financial improprieties, customer deception, bribery or improper influence, regulatory violations, corruption, or undisclosed conflict of interest. One study found that 62 percent of all companies experienced a "significant or major" integrity continuity disruption between 1986 and 1996 (Gottlieb & Sanzgirli, 1996).

These scandals can disrupt business operations; negatively affect performance; produce entangling regulatory scrutiny or litigation; create labor problems; damage reputation, image and brand; and produce criminal charges against senior management or top executives. Sentencing guidelines, such as those of the Uniform Federal Sentencing Guidelines for Organizations hold executives and senior management accountable by instructing judges to consider organizational efforts to plan, train, and implement policies to mitigate; enact full-disclosure efforts; and cooperate with authorities. One important conclusion emerges from these expectations: Corporations are increasingly held accountable for their efforts to educate, train, and enable employees to act ethically, legally, and with integrity in the performance of their duties.

It is important to note that unethical scandals can occur in any organization despite what employees believe about their institution. Many individuals take on the mantra that "it couldn't happen to me." However, ethical breakdowns happen consistently, and they strike even the most prosperous of companies. It is essential that companies take preemptive strikes at issues that question integrity in order to protect themselves. Being overly prepared is far more beneficial than refusing to implement integrity policies and, thus, having to face the consequences of their actions. Preserving integrity in the workplace involves shaping employees' perceptions of expected ethical conduct and the punishments that will arise if they refuse to be acknowledged.

Preventing Ethical Misconduct Disasters. While it is impossible to prevent 100 percent of ethical issues that arise, it is possible to take steps that can limit the frequency of such episodes and that might very well limit the expansion of an ethics issue into a major scandal. Furthermore, these steps are also useful in helping mitigate the damage and impact of an ethical disaster when—not if—they occur.

Integrity Continuity Assessment. Integrity continuity considerations are commonly ignored by senior management, particularly at the chief executive level. Research has found that 60 percent of chief executives and boards of directors failed to engage in integrity continuity planning discussions or to include such considerations in strategic planning (Taub, 2002). Furthermore, 57 percent of companies "have never" incorporated integrity continuity planning at the strategic executive or board level (Taub, 2002). As mentioned previously, most businesses do not regularly or systematically assess their vulnerability to ethical misconduct risks, and few take proactive steps to ensure integrity continuity. Most companies do not include ethics aspects in their annual performance appraisal criteria, nor do most companies conduct ethical compliance audits.

Assess and Monitor Employee Behavior and Decisions. Formal codes of ethical conduct alone are insufficient to prevent scandals or ensure integrity continuity. Active ethics efforts are the key for integrity continuity. Ethical concerns must be regarded similarly to other business disruption/resumption concerns. In most cases, ethical disasters involve employees who failed to follow their own internal corporate policies and guidelines. Rarely, if ever, does a scandal arise from a single individual who was acting independently from the larger system, culture, process, and reinforcement of expectation norms, all of which tend to facilitate, encourage, or reward the unethical behavior. These risks can be minimized by assessing employee conduct and integrity of the organizational culture (including measuring employees' knowledge and familiarity with polices, rules, and protocol); conducting effective, active ethical conduct training; and evaluating the effectiveness of ethical codes of conduct and training efforts.

It is imperative that organizations monitor and audit employee conduct in order to have a realistic idea of the kinds of behaviors and decisions that occur within the company. Start with a comprehensive Ethical Conduct Audit© (ECA) to get a snapshot of the current enactment of integrity across an organization. It would also be useful to review past decision-making processes and situations where there is vulnerability to integrity lapses. Ethical training policies should serve to train employees to recognize and make ethical decisions. Organizations should review participation efforts to ensure that all behaviors and decisions are being conducted ethically. Finally, it is important to create and maintain whistle-blower channels, policies, and protections for those who report unethical conduct.

Maintain Ongoing Proactive Integrity Continuity Management. It is essential to maintain effective programs that have been established. It is easy to create programs and then fail to assess, maintain, or support them on an on-going basis. Managing integrity continuity requires ongoing commitment and proactive attention to maintain the processes and to achieve, continuously, the ethical goals. Senior management should insist on systems and processes that recognize and reward acts of integrity and ethical decisions. The overall policies and procedures should naturally lead to conditions where employees routinely abide by the desired ethical practices. The processes of appraisal should inherently include ethical assessment, and managers should enforce disciplinary policies consistently and fairly. Managing integrity continuity requires on-going commitment and proactive attention to maintain the processes and continuously achieve the ethical goals. Businesses should create and maintain a supportive climate for ethical conduct by recognizing and rewarding acts of integrity and ethical decisions.

It is also important to conduct organizational transformation (OT) activities, training, and development programs to build a corporate culture of integrity. Managers should anticipate potential threats to integrity continuity and take proactive measures to mitigate problems. Management should develop and practice plans for reacting and responding to the discovery of unethical behavior. Finally, it is important to have contingency plans for handling issues that might potentially become major scandals and disruptions to ongoing operations.

Commitment to Integrity Continuity. To survive crises of integrity, a company must make ethics policies fundamental to the core of their organization. The company should be committed to enforcing these policies and proactively managing integrity. In order to reinforce the importance of these policies, corporations should expect, reward, and encourage ethical behavior in their employees. Specifically, employees should not be required to bend the rules to effectively perform their job tasks. It should be consistently reinforced that employees can and should act in ethical ways to complete their work. Managers should never, even inadvertently, reward an employee who achieved a goal at the expense of integrity. Ethical decisions should be ingrained in the minds of the employees as facilitated by culture, attitudes, and actions of the other employees and managers. Employees should be trained to make integrity a work habit.

To survive crises of integrity, a company must maintain a fundamental commitment to the strategic goals of proactively managing integrity. Post-crisis analysis of many of the major ethical scandals has revealed that employees were frequently motivated by formal and informal reward systems and explicit and implicit expectations of management, or that they believed themselves to be acting consistently with the behavior of other employees in the organization. Far too many corporate scandals have occurred because the organization (culture, policies, or reward system) was an enabler of the employees' unethical behavior.

Integrity is a process and way of working. Employees can be trained to make integrity a work habit. Ethical conduct is a dividend-paying, long-term investment—not an expense—and is a fundamental breach of obligation to stakeholders.

Establish Explicit Ethical Goals and Criteria. It is important to define clear and explicit goals and behavioral criteria when constructing integrity continuity plans. Codes should be detailed as well as applicable. Expectations of individuals, whether legal or ethical, should be stated clearly to prevent any confusion. It is important that these criteria are written. They should include examples in order to make the goals applicable to real life. These specific situational examples provide a clear, unambiguous, and easy-to-follow normative standard to guide employees as to what is or is not considered behavior or decisions consistent with the company's expectations. The more descriptive the types of examples are, the more readily employees will be able to recall them and apply them when confronted with ethical decisions in the workplace. All of the ethical codes should be put into writing, distributed to employees, and reinforced consistently.

It is also important to ensure consistency between reward systems and ethical integrity. There should be an alignment of promotion, compensation, and rewards with actions that demonstrate ethical behavior. Many companies unintentionally have pay, reward, recognition, and evaluation criteria that actually encourage or reward unethical behavior. It is vital to have an alignment of all incentives with the ethical goals of the organization.

One useful recommendation that helps reduce the risk of ethical misconduct disasters is the simple step of adding ethical action and integrity behavior as an explicit evaluation item on the annual or semi-annual employee appraisal evaluation criteria. This signals that the issue is truly important to the company. This also provides an explicit incentive to act ethically or with integrity and the more so, the greater the reward for ethical conduct. This leads to the next important aspect, which is that an organization must demonstrate a top-level commitment to ethical goals and standards.

Demonstrate Commitment to Those Ethical Goals and Criteria. Simply paying lip service to integrity and ethical conduct is not enough to avoid ethical issues. For many senior managers, having a Code of Conduct document prepared and distributed across the organization is assumed to have fulfilled their obligation to demonstrate commitment to such principles. Unfortunately, this has not proven to be the case in many instances. Some of the most significant ethical misconduct disasters of the past decade occurred in companies that had established and published ethical codes of conduct and ethical goals and guidelines.

While having a written code that explicitly defines ethical expectations is important, it is also necessary to illustrate to your workforce that the company takes ethics in the workplace seriously and is committed to expecting employees to meet and exceed such standards. It must be made clear to employees that any statements regarding ethical conduct are serious and not just lip service. It is important for employers to make it clear that they are committed to integrity—not just for moral purposes but for strategic goals as well. It is beneficial to create a position known as a corporate ethics officer, whose job is to manage a strategic integrity plan as well as to signal commitment. A rewards system should be put into place to establish indicators of integrity in addition to other measurements of productivity. It is important to consistently reward integrity and to make sure that these instances are known throughout the organization. It is equally important to have a clear and efficient disciplinary process for lapses of integrity.

Communicate Ethical Expectations and Train Workforce to Enact. Organizations need to create integrity expectations and ethical criteria, and the employees should be aware of these expectations. An ethical training program should be implemented as part of the strategic commitment to integrity continuity management. It is important for employees to recognize situations in which integrity plays a main role, and it is crucial for them to understand how to avoid lapses of integrity. Specific situations and decisions that are faced on a daily basis should be used in order to illustrate how ethical principles can be implemented in employees' jobs.

Key Questions

The contemporary track record highlights the necessity for companies to begin to alter their mindset concerning integrity continuity planning. This change must be led from the top. Every senior manager, executive, or corporate board member must begin to ask the key questions about integrity continuity. Incorporating ethics as a strategic high-level commitment serves to ensure the protection of employees, stakeholders, executives and board members themselves, and the company as a whole. A beginning step to this process is the assessment of the unique integrity risks as they pertain to a specific company. Questions to assist in this analysis include the following:

- On what criteria do you base confidence in your company's integrity continuity?
- Do all company personnel know how to act or behave ethically and appropriately in all situations and contexts?
- Do employees know the rules for each situation that may arise?
- How does the company know the employees have this information?
- How do you know, and how certain are you, that your employees are acting ethically?
- How do you define the desired ethical behavior?
- How do you implement the criteria so they will be accepted and followed by employees?
- Do your employees know where to turn for resources and support for acting ethically?
- Are your employees aware of how to report ethical issues?
- Does your performance appraisal process specifically reward ethical behavior and punish unethical behavior?

Based on the answers to these questions, policies and codes can be formulated to fit the company's needs. These basic criteria alone are not sufficient to prevent scandals or ensure integrity continuity, however. It is essential that ethical practices are actively carried out within organizations and that policies are consistently upheld. Ethical disasters usually result from employees' failure to maintain ethical standards; rarely do scandals arise in settings where integrity continuity plans have been put into place and practiced on a daily basis.

In order to minimize risks of ethical disasters, a number of factors should be considered. Some of these include the assessment of employee conduct, integrity of the organizational culture, conducting effective and active ethical conduct training, and evaluating the effectiveness of ethical codes of conduct and training efforts. An approach that has been valued highly in its ability to effectively assess a company's strengths, weaknesses, opportunities, and threats is the ECA. This assessment can gather information, establish reporting channels, assess culture, examine Perceived Reward Systems, be alert to warning signs, and identify patterns of potential misconduct. The benefits of this system include its provision of insight into both legal compliance behaviors as well as into the ethical reasoning and decision-making that is often difficult to observe.

Ethical Conduct Audit© (ECA)

The Ethical Conduct Audit© (ECA) is one approach that has been used effectively to assess a company's strengths, weaknesses, opportunities, and threats (Chandler, 2000). Such an assessment process can provide insight into both legal compliance behaviors as well as into the ethical reasoning and decision-making that is often difficult to see with unfocused or casual observation.

The goals for the ECA include identifying relevant ethical issues confronting the company or agency; assessing current behavioral practices and benchmarking those practices; ascertaining compliance with applicable internal and external rules, laws, and regulations; reviewing the role and sufficiency of current ethical policies, codes of conduct, and ethics training; monitoring the impact of new initiatives or programs in the organization; and closing the loop from changes implemented based on precious ECAs.

The ECA is a multi-method empirical approach utilizing independent measures to obtain a snapshot of current practices, attitudes, beliefs, and norms among employees. Typically, a combination of questionnaires and interviews is administered. In addition, the ECA usually includes direct observation of behaviors, focus group sessions, and a review of communication artifacts, including whistle-blower tips and confidential reports. The ECA process is adapted to the unique needs of a specific organization.

Frequently, the ECA can help identify essential or core ethical vulnerabilities for a specific company or agency. The audit can be a tool to help identify potential risk vulnerabilities (e.g., conflict of interests, discriminatory practices, incentives for unethical behavior, etc.). In addition, the audit can provide insight into how well the workforces have knowledge of procedures, codes of conduct, regulations, and policies along with some sense of adherence and compliance with those policies. The ECA can help identify warning signs of ethics risks before they become serious problems. In some cases, the ECA is a mechanism for reporting violations and learning about ethical problems and issues before they become full-blown ethical misconduct disasters.

One unique aspect of the ECA is that it also focuses on ethical processes in decision-making. It may help provide an early warning for potential unintended consequences of the various decision processes.

The RADAR Model

If an ethical scandal does occur, the response must be quick and decisive. Reaction to instances of unethical behavior must be immediate. It is always important to follow procedures consistently and fairly. It is also important to document all aspects of the efforts for integrity management as well as all steps for responding to episodes of ethical misconduct, followed by immediate action. Rest assured that every step taken, or not taken, will be closely watched by other employees and eventually by the media and the public at large. If employees know their internal procedures and policies as well as the legal and regulatory requirements, follow disciplinary policy consistently and swiftly, contact and cooperate with law enforcement (when appropriate), and review stakeholder agendas, then possibly the company may emerge gracefully from the disaster.

Dr. Linda Ferrell offers a quick acronym (called the RADAR model (Brewer et al., 2006) to summarize an approach to basic steps for EMD management and survival. Ferrell summarizes the approach as the organizational RADAR to identify and deal with ethical disasters. The five terms for the RADAR model are recognize, avoid, discover, answer, and recover. The RADAR model points toward five distinctive aspects of mitigating the risks of ethical misconduct disaster.

Recognize. There must be a strategic level effort to define the ethical issues. This should incorporate threat forecasting, risk assessment, and demonstrated anticipation to recognize the dangers of ethical lapses and their resulting scandals. This should also include the types of comprehensive risk assessment and ethical assessment discussed earlier in this chapter.

The most important first step is to admit that ethical issues are a problem and that the threat of an ethical misconduct disaster is quite real and serious.

Avoid. To avoid EMD risks, it is essential to develop proactive ethical leadership in the organization. Such leadership should demonstrate and role model the goals for corporate integrity. Ethical leadership must be consistent and reinforcing with ethics policies, rules, codes of conduct, and practices that help craft an overall ethical culture and climate. This includes effective ethics/compliance programs, ethics training initiatives (including ethical decision making); development of skills for handling difficult situations, and reinforcement for acting with integrity.

Information and expectations for ethical conduct must be communicated effectively in a wide variety of formats, methods, and channels to ensure that it has an impact on everyone in the organization. It is essential to develop substantial and comprehensive ethics and compliance programs to help avoid EMDs.

Discover. Mitigating an EMD requires vigilance and the capacity to discover threats and episodes early. This includes scanning for threats and problems, listening to feedback about ethical issues, monitoring behavior, actively communicating with employees, and incorporating both formal and informal feedback systems.

The deployment of a confidential reporting system as well as an active on-going assessment, including periodic Ethical Conduct Audits©, to discover the warning signs of impending ethical issues is an important strategy. Discovering ethical issues sufficiently early to intervene as well as a timely, adequate response can help reduce the risks of major scandals and better ensure that one can survive if and when they do occur.

Answer. When ethical breakdowns occur it is critical that management quickly responds to the uncovered situation in order to prevent an issue from growing into a scandal or to keep a scandal from becoming an ethical misconduct disaster. This should be integrated with an organization's ongoing oversight, integrated systems, rapid actions, and ethics program development. However, it also requires fast and decisive actions to nip in the bud emerging ethical concerns.

Answers to ethical concerns are both practical and symbolic gestures, which may make the difference between a controllable ethical event of which the organization is proud of how the situation was handled versus an out-of-control event that damages worker morale, market value, brand, corporate image, and reputation—and may lead to regulatory fines and criminal sanctions. It is useful to have a series of integrated systems in place, which will help detect and discover ethical concerns but also guide the organization through specific response procedures and processes. It is important for any answer to ethical issues that arise to be prompt and appropriate. This may only be possible if protocols, including reporting and disclosure rules, have been established well in advance of any questionable behavior or troubling situation.

Recover. The possibility of recovering from a major ethical misconduct disaster may depend on how well an organization has prepared for the process of any scandal. Some ethics scandals can consume and overwhelm an entire company. Others can actually lead to stronger levels of commitment to integrity and ethical ideals. The reassuring news is that many companies that are rocked by such scandals can eventually earn back the public's trust and their reputation in the marketplace.

Major ethics scandals can damage productivity, hurt shareholder value, and suppress consumer confidence in a company. The term *disaster* may be an understatement to describe the total and cumulative impact of a major ethics scandal on an organization.

While attempting to put the pieces back together again in the aftermath of a major ethical misconduct disaster, a company will find that people are often demoralized and unmotivated; the organization feels distracted and in disarray; and that routine operations are disrupted. Legal, regulatory, and criminal proceedings may be underway and can drag on for months and years to come. Key people may be away from their tasks dealing with such investigations and proceedings or even perhaps serving prison time.

Even if not so distracted by official processes, many employees will have low levels of motivation and morale. A lot of really good employees will leave companies rocked by scandals, making the recovery process all the more difficult for those who remain.

The road to recovery requires both advance planning and diligence to implement. Organizations and companies should have a road map to help rebuild trust in employees, investors, and customers. They will need to take strategic steps to restore image, stabilize the situation, and resume productivity. A comprehensive and coordinated communication plan that includes active listening is essential to recover.

Conclusion

It is amazing to hear a top corporate executive make a statement that ethical scandals "couldn't happen to us." Those sentiments are intrinsically related to the aftermath statement, "I never thought this would happen to us." The reality is that all of the common justifications for ignoring integrity continuity planning are based on unsubstantiated (no systematic psychometric assessment that would provide a substantiated level of confidence) trust in unmanaged human nature and in ignoring the systemic factors that give rise to ethical disasters. Managing integrity requires more strategic planning and enactment beyond hiring good, moral people. Even systematically hiring only employees with the highest levels of morals and ethics is no sure-fire method for preventing a major scandal (i.e., consider the significant scandals questioning conduct of some priests within the Roman

Catholic Church). Neither does having a detailed written statement of ethics or specific documented policies provide a sure-fire method to prevent such disasters. For example, the final 65-page Enron corporate Code of Ethics was written in 2000 and was intended to help guide employees for "conducting the business affairs…in accordance with all applicable laws and in a moral and honest manner" (Enron Corporation, 2000).

In addition, one cannot adequately avoid these scandals by delegating such planning to subordinates. Managing integrity continuity requires that organizational standards and expectations be clarified and made a top priority. Such standards must be used in the formation of formal and informal processes such as compliance policies, performance appraisals, and employee training and development efforts. These efforts should be an implemented part of an active and on-going specific ethics training program to ensure employee alignment with the goals as well as providing them with the knowledge, decision-making skills, and abilities to recognize ethical danger issues and how to successfully navigate through them. The core focus on ethics should be meshed with all of the decision-making processes that occur at various levels and functions in the organization. To ensure integrity continuity, it should also include regular and periodic assessments of employee behavior and awareness of potential ethical issues as early as possible.

All of these efforts should be considerations at the highest levels of management. Integrity continuity planning must occur as part of the strategic planning process. It must be integrated with decisions on development, transformation, goal setting, prioritization, and regulatory compliance issues. The fact is that every organization is vulnerable to risk of an ethical misconduct disaster, and how effectively they recognize these risks, anticipate these challenges, and move to prevent such scandals may ensure their survival.

References

Brewer, L., Chandler, R. C., & Ferrell, O. C. (2006). *Managing risks for corporate integrity: How to survive an ethical misconduct disaster*. Mason, OH: Thomson.

Chandler, R. C., & Wallace, J. D. (2001). Brief results of the Pepperdine University ethical misconduct disaster recovery preparedness survey. *Disaster Recovery Journal*, *14*(3), 21–22.

Chandler, R. C. (2000). Managing ethical and regulatory compliance contingencies: Planning and training guidelines. *Contingency planning and management 2000 proceedings*. Flemington, NJ: Witter.

Enron Corporation. (2000). *Enron code of ethics*.

Gottlieb, J. A., & Sanzgiri, J. (1996). Towards an ethical dimension of decision making in organizations. *Journal of Business Ethics*, *15*(12), 1275–1285.

Integrity and Ethics in Public Administration: Polls and Surveys. (1998). *Public management*. Washington, D.C.: Public Management.

Josephson, M. (2002). *Report card 2002: The ethics of American youth*. Los Angeles, CA: Josephson Institute of Ethics. 2002. http://www.josephsoninstitute.org/Survey2002/Report-Card_data-tables.pdf

Lazere, C. (1997, April). Ethically challenged: Teaching ethics is required but schools have wide latitude in how they do it. *CFO Magazine*.

LeClair, D. T., Ferrell O. C., & J. P. Fraedrich. (1998). *Integrity management: A guide to managing legal and ethical issues in the workplace*. Tampa, FL: University of Tampa Press.

Taub, S. (2002, June). Crisis of ethics. Retrieved from http://www.cfo.com/article.cfm/3005220?f=search

Walker Information and Hudson Institute. (1999). *Workforce 2020*, Indianapolis, Indiana: Walker Information and Hudson Institute.

Section 2

Integrating Risk Communication into Pre-crisis Planning

Chapter 6

Pre-crisis Management and Communication: Slippery Steps or Solid Footing?

Robert Heath & Judy Sultan

This chapter reinforces the need for crisis planning, preparation, and management that set the stage for effective crisis prevention and response. The authors identify challenges faced by private sector organizations, government agencies, and non-profit organizations where executives fail their organization and stakeholders/stakeseekers. In essence, the chapter stresses the need for individuals and all types of organizations to make pre-crisis management and communication policies along with good implementation strategies a priority.

Introduction

As a father and husband, the senior author advised his wife and daughter to keep an eye on the gas gauge of each of the cars they drove. He purchased jumper cables for each car in the event they might need someone to help start their car if its battery failed. Once cell phones became readily available, each was armed with one. He had automobile insurance and AAA membership. Does this mean that he was a compulsive worrier or a "pre-crisis" manager trying to be ready for various risks that might manifest themselves? Although those options are not necessarily binary, he hopes that you answer the latter.

That example frames the topic of this book: pre-crisis management and communication. The purpose of this chapter is to set a foundation for the discussions in other chapters. In part to accomplish that goal, this chapter draws on the BP (British Petroleum) oil production crisis and includes a case study based on the marred athletic career of Michael Vick, who arguably lost a fortune because he did not manage his professional career and public reputation/brand equity properly—nor did the National Football League (NFL) offer proper pre-crisis prevention and response assistance.

Crisis discussions often feature challenges faced by private sector organizations. Indeed, there are many iconic cases in business but also in government and non-profit organizations where executives fail their organizations and its stakeholders/stakeseekers. Presidents of the United States' actions mar their reputations and put policy positions at risk (or cause their failure). Members of Congress become involved in ethics scandals. Non-profit organizations blunder. So, organizations of all types need pre-crisis management and communication policies and their implementation.

Chapters following this one offer specific insights based on research and theory that can assist organizations toward enlightened and reflective management as a first step to not only prepare for a crisis should (or when) it occur but also to work diligently and wisely to prevent crises—and mitigate them if they occur.

As this chapter was being written, oil was flowing after the Deepwater Horizon drilling rig explosion, a well drilled in the Macondo Prospect in the Gulf of Mexico under BP management. After more than three months, the well was capped and killed. Every stakeholder had asked, *when will the flow end and how extensive will the damage be? In addition to BP, what other organizations, individuals, animals, and plant life suffered crisis: partners in the well, oil field service companies, fishermen, persons and businesses dependent on tourism, wildlife, and plants?* That last question reminds us to not focus only on one organization (BP in this case) as we address crisis impact and the dialogue that follows an event or revelation. Crisis on the part of one person or organization has a way of resulting from and creating crises for others.

Among the lessons re-learned from BP's crisis is that planning must consider wisely and proactively the likelihood of various risks that can manifest in ways that require prevention, mitigation, and response during very deep water exploration and production. The case revealed that BP did not have a plan that worked to cap the well quickly and properly. The blowout preventer failed, perhaps because of faulty design, maintenance, or operation. Typical methods to control runaway wells either failed, were inadequate, or perhaps did additional environmental damage (such as dispersants used to break up the oil). The number of management and operation mistakes or failures (miscalculations), starting during the final production steps that led up to the blowout, played out in real time on television. BP's communication plan failed or was inadequate for the challenge.

It is wise from the outset to emphasize that crisis usually does not result from "communication breakdown" but from management and operations blunders. Thus, the first step toward pre-crisis communication is not communication planning but reflective and vigilant management. We must start our discussion by noting that performance guidelines that are knowable and crises that are preventable become a standard of management excellence. We become interested in organizational culture, perception of the role of the organization in society, and the ability of management to understand and be willing to meet or exceed performance expectations. We are as interested in systems, functions, structures, and technical competence as we are in message design and delivery.

Pre-crisis efforts are not limited to developing, practicing, and refining a communication plan. Having a good plan is important, as having automobile insurance is a solid first step toward that moment when, in a test of physics, two drivers seek to occupy the same physical space at the same time. Insurance is a management step toward crisis response, but safe driving attitudes and defensive driving skills are essential to reduce the likelihood that a crisis will occur. If it does happen, sound management can help mitigate the event and its impact on various risk bearers.

We end this opening statement and lead into the body of our chapter by noting that pre-crisis management and communication require certain reflective attitudes and behaviors. In a world where apology and saying *I'm sorry* often can be seen as a reasonable response (as may blameplacing), in fact such behavior often masks the tragedy of mining practices and emergency response errors that lead to loss of life. Ability to stand and say, "We erred in the design and construction of levees," does not make whole a city that suffered a devastating flood, nor does saying, "We were assured that the cementing stage, although difficult, was developing at an acceptable level of risk—and if successful would bring this well into production."

Over the past decades, following Johnson & Johnson's Tylenol recovery, analysis often obscured the multi-disciplinary challenges of crisis. Discussion focuses on what to say and do once something goes bump in the dark. As crisis has been studied by management practitioners and scholars, it has also been examined by technical experts such as engineers and other scientists. It is in their training to prevent and mitigate negative consequences, through—for instance, better designed and maintained operating systems. Lots of constructive discussion can occur between risk managers in finance, accounting, law, and other disciplines that couple with the discussion by management and technical experts.

In the mix, we have communication experts. Authors in this book fit into that category, but every statement they make has to presume that they cannot responsibly discuss pre-crisis communication without seeing it as interdisciplinary. Also, the members of other disciplines cannot act reflectively and responsibly by assuming that if they have a good communication plan, it can explain anything away, even ostensibly irresponsible decisions and illegal actions.

With these overview thoughts in mind, we discuss key topics to challenge and frame the chapters that follow. The ultimate outcome goal is to bring together thinking that can help mitigate impact on risk bearers and make society more fully functioning by the wise management of crisis.

Forethought and Planning: A Troika of Pre-crisis, Crisis, and Post-crisis

Our overview features broadly a narrative perspective to pre-crisis planning, prevention, mitigation, training, and response (see especially, Heath, 2004a). It is part of the archetypal advice typical of our lives: If you fall and hurt yourself, don't say I didn't warn you. Look before you leap. Imagine that every stick might be a copperhead, and one is hiding in every pile of leaves. Think where your fingers are before you operate a miter saw. Sound the signal before detonating the explosion. Tag (cross or double tag) a valve before (or after) opening it. Holler "Heads up!" before throwing a ball to someone who is not watching. Be prepared to holler "Fore!" on the golf course. Consider every step of an operation before beginning as a team to handle high-voltage lines. Back up computer files. These statements suggest how much of our lives depends on pre-crisis planning, training, management, and reflection. These contextual narratives recur as we go about routine activities.

Pre-crisis, crisis, and post-crisis stages have become featured aspects of crisis analysis (See Coombs, 2009a; see also Millar & Heath, 2004). How well we understand these stages as discrete but interdependent events (points of analysis) and know what each requires (both as prevention and response) can advance the theory, research, and practice. Pre-crisis communication can ask *what can be said and done to reduce the likelihood of a crisis and mitigate its harm if it occurs?* This, for instance, is a key aspect of effective community relations crisis communication where high risk companies, such as petrochemical facilities, work to prevent crisis but also communicate with area residents to mitigate damage if or when a crisis/emergency occurs. Similar logics apply to experts who predict and alert residents to severe storm warnings as well as respond when such events occur.

Just focusing on the crisis event can be dysfunctional. As Coombs (2009a) reasoned, pre-crisis should embrace concerns for prevention and preparation. The crisis stage is concerned with response: process and content. The post-crisis phase gives various voices and managements the opportunity for follow-up communication—perhaps, for instance, offering the lessons learned from the crisis that can reduce the likelihood of recurrence, mitigate it if it does, prevent recurrence, and prepare stakeholders for recurrence. Working to understand effective post-crisis response, Ulmer, Sellnow, and Seeger (2009) reasoned that this is a time for the organization to renew itself. In one sense, that option suggests that the organization cannot simply seek to return to business as normal, but it must position itself to become different and better. Jaques (2009) reasoned that issues management is challenged and empowered as post-crisis response. But issues management can do a lot prior to a crisis—especially prevention and mitigation (Heath & Palenchar, 2009).

Emergency management, including hurricane research, tracking, and warning, is important for pre-crisis response. But it does not start or stop with preparing and reporting on weather conditions. Emergency response (individual actions) and communication planning and response (including evacuation) are important considerations. Safety training requires that you ask: What do you do, and how empowered are you as an employee who can stop an assembly line—or who finds an unknown substance (or a leaking valve, etc.)?

Pre-crisis communication is a form of applied communication based on theory and research—as well as ethical and tested best practices. As much as we might want to feature deontological philosophy in discussing the ethics of pre-crisis, at the management table, a heavy dose of pragmatism nevertheless needs to be injected as well by asking, "Whoever wants to be held responsible for making or not making that decision, please raise your hand." That challenge can go a long way toward grounding pre-crisis planning and management in reality.

The Deepwater Horizon crisis reminds us that a bad news day is not a crisis. Bad news events often do not last beyond one or two days of news cycle; they are unlikely to seriously affect the reputation, legitimacy, and even viability of an organization. The myopia of only focusing on the crisis suffered by one organization (and of minimal impact) began with cases where heavy media scrutiny defined crisis. The original interest in crisis management and communication arose because some company was on the hot seat for what it did or did not do. Today, however, we realize that crises reach out in complex ways so that many different crises combine as challenges that constitute facets of a single crisis. They have various impacts, including the end of the organization as it was before the crisis. Think Enron, WorldCom, and Union Carbide.

The BP case reinforces our desire to take a narrative approach to communication, especially crisis management and communication (Heath, 2004b). The BP narrative became visible when media alerted the public to the fact that a drilling vessel had caught fire and sunk, and lives had been lost—or key personnel were missing. At that moment it was a Transocean and BP narrative. Slowly, it became a crisis narrative for the company that owned the drilling vessel, the various oilfield services companies that provided equipment and expertise, and the men and women who made decisions (however large or small) that led to the fire and uncontrolled flow of crude oil. The original narrative offered by BP soon collided with narratives put forth by other voices: scientists, government officials, members of the drilling crew and those on service vessels near the rig, and local business and government leaders in Louisiana. Within a week after the triggering event, many narratives were competing for media space and credibility. As is true of all narratives, those that seem closest to fact, fidelity, and are most coherent sustain themselves best without reinterpretation and reassertion (Fisher, 1987). And media as well as governmental scrutiny slowly peeled back the time frame to help understand what had led up to and caused the crisis.

Even though the crisis began at the point of media attention, the pre-crisis phase had gone on for a long time. It included technical planning, equipment design and testing, experience with equipment and operations, and other key aspects of oil discovery and production. Someone in the decision stream had to make decisions as to how and whether to proceed with high-risk actions. Did they know that other wells under similar pressure and reaction profiles had caused problems? Did they know that other companies had stopped drilling and production as well as capped wells that were too risky? Was this well known for the difficulty and challenges of bringing it to production? Was the production team vigilant in terms of maintenance of critical blowout prevention equipment that was to be designed to prevent such occurrences? Was government oversight through the Minerals Management Service (MMS) sufficient? Are the rumors true that the MMS personnel were lax, biased to the industry, understaffed to meet their mission, and often willing to allow operators to complete oversight paperwork? Were the response documents badly drafted, merely as a formality rather than a precaution?

Over time, narratives featured the magnitude (a point of dispute) of the amount of oil coming out of the pipe, the oil on the surface as it migrated onto shore that suspended in one or more plumes below the surface of the water, the currents that would carry oil far from the spot of the event, and the impact. Impact centered on persons whose livelihood depended on fishing, dredging for oysters, and tourism. Those industries were not merely limited to operators of shrimp boats but the persons who processed the food—and cooked it. Like the ripples that emanate from a rock thrown into a pond, slowly but inevitably the ripples reached well beyond one splash. Then, the crisis seemed to be best characterized as many pebbles thrown into a pond; the ripples rolled out, intersecting and affecting one another.

Many premises that guide crisis research and best practices seem relevant to this event. One is this: "Careful planning increases the likelihood that crises will not last as long and be as damaging as they would if less careful preparation had been made" (Heath, 2004a, p. 35). BP, like other drilling companies as well as oilfield technologies/services companies, has worked hard to improve drilling and production performance, both in terms of return on investment and safety. They aspire to zero environmental impact because that's what their stakeholders expect. They realize the wrath of government and other stakeholders will descend on them if errors are made and damage occurs. They had developed knowledge and skills to not only reduce the likelihood of a major release but also how to respond if one should occur.

However much they had planned to respond, planning failed because the response failed. The company looked inept and suffered a giant blow to its credibility, which rippled onto the industry. As is often the case, a single event becomes the narrative that overshadows many success stories. Planning failed, perhaps because shortcuts were taken at crucial stages as the company worked to bring a difficult well into production. As the cultural archetype says: Haste makes waste. The entire industry received a black eye, was stopped from operation for a while, and incurred regulatory backlash. The most dramatic failures were BP's attempts to stop the flow from the well—producing the narrative of clumsy, unskilled, and foolish responses. Comedians feasted on its struggle.

This was true of the *Exxon Valdez*: the public mind only remembers the one tanker that spilled oil in Prince William Sound—not the thousands of safe trips that best characterize the skill of the industry. In the same way, BP's Deepwater Horizon well will become iconic and shape public opinion and policy toward deepwater drilling and production in the Gulf of Mexico, one among the thousands of completed wells. This brings us to two additional propositions. First, we emphasize "what if":

> Organizations that are more self-reflective are likely to create and implement a more successful crisis response and communication plan. Such organizations are more likely to develop multiple options and achieve requisite variety. (Heath, 2004a, p. 35)

How reflective was BP in planning and preparing for a major drilling failure? And, how long will it be before another CEO of a targeted company publicly fumes that he wants a crisis to end so he can have his life back? This lack of reflection suggests the wisdom of another proposition:

> Organizations are likely to suffer less damage from crisis when they are prepared to repair relationships and demonstrate commitment to promote and protect their stakeholders' interests. (Heath, 2004a, p. 35)

For at least three decades, interest in organizational crisis has created a cottage industry for experts keen to hone management practices. It is even a key aspect of issues management insofar as issues (for example, an unchecked concern such as discrimination in the workplace) lead to crisis (activists confront an organization, sue, and demand reform). And attack crisis (for example, malevolent acts of terror such as 9/11) leads to issues (policy changes in airport security, states coordinating issuance of drivers' licenses, etc.) (see Heath & Palenchar, 2009). The management reasoning for this interest is simple: Crisis costs money, which offers the incentive to avoid, mitigate, and respond in ways that best protect capital and human resources and generically "reputation" or "image" that some feature as the essence of effective crisis response. Damaged reputation can offend businesses' customers, non-profits' donors, and legislators who provide tax revenue for government agencies. The ultimate recovery objective features the integrity and legitimacy of the organization, as managed resources, through various disciplines, including public relations.

Development of a Discipline

Crisis management, prevention, mitigation, and communication have become standard course offerings at colleges and universities, a consultancy specialty, and the topic of endless training and planning sessions at various levels with large and small organizations. It is carefully orchestrated by management teams and often produces tensions between general counsel and public relations experts. Management is never separate from crisis prevention, mitigation, and response. The best managed crisis is the one that never occurs or, if it does, quickly ends because of sound management.

Such advice is not only important for businesses but also non-profits and government agencies. Their crises often result from mismanagement of funds or violation of the organization's mission and vision. Various non-profits are specifically and strategically positioned (mission and vision) to respond to crisis. The Red Cross is one of the most obvious organizations of this type (Schwartz, 2010). Likewise, the Federal Emergency Management Agency (FEMA) is a government agency designed to be a crisis responder. Hurricane Katrina created a crisis for FEMA, which suffered crisis by its clumsy efforts to minimize or mitigate the crisis experienced by other organizations and myriad individuals.

Also, celebrities play beyond the edge of crisis control by flaunting social convention to increase publicity value. Given that they are celebrities (same seems true for public officials), they may come to believe they are either above the law (or social convention) or allowed (even encouraged) to play by different standards of acceptable conduct. And, the violation of those standards, which could harm the reputation of the ordinary citizen, may enhance celebrities' market values.

These twists and turns motivate academic and practitioner investigations of how ethically sound and strategically smart responses should be employed. Academic research opportunities abound, no further away than today's news. Senior managers in major organizations collect case studies to introduce into executive discussions a higher standard of ethics—corporate social responsibility—and need for reflection and sound planning. Practitioners have a revenue bird's nest on the ground. Students thrill at the drama of crisis and thirst to be crisis responders. Younger practitioners often believe they have "earned their stripes" when they get to respond during a crisis. Cost of badly managed crises can even outweigh the often unfortunate countervailing influence of general counsel to say and do as little as possible, lest it be held against you in court.

As we navigate the pedagogy, research, and best practices relevant to crisis, we pause periodically to assess seriously where the literature and best practices are, where they should go, and how we are going to get there. In such discussions we are increasingly mindful of the twin pillars of process and content/meaning. Various processes—before, during, and after a crisis—can affect how it plays out. Meaning matters; interpretations and evaluations of crises are socially constructed and rhetorically challenging. Discourse analysis can shed light onto the texts that lead to, surface during, collide, and become resolved after a crisis. Often such analysis has keen insights by hindsight. Discussion of pre-crisis requirements helps us to reorient our thinking on this matter.

Crisis by Any Name

What is a crisis, and therefore what is pre-crisis? Heath and Millar (2004) listed approximately 20 definitions of crisis. Most definitions feature newsworthy events. For that reason pre-crisis can be slighted in the discussion and conceptualization.

Some definitions feature a mistake or dramatic turning point in the history of an organization. Others focus on the need for management efforts beyond normal or routine procedures. Some emphasize stress, inadequate control, uncertainty, violation of laws or ethics, and other malfeasance. Some point to weak preparation and inadequate prevention—as well as the need for crisis communication planning, training, personnel role assignment, drills, and other strategic and tactical options, including drafted or templated messages to be used when the crisis occurs because they will—and do. Holding statements for this reason, as well as blocked Web homepages relevant to various crises, are held for such eventualities.

With a focus on some critical event or condition, one can define crisis as *a risk manifested.* As such, organizations can be defined and evaluated by the quality of their risk management, which, if it cannot prevent a crisis, it can at least understand the conditions and requirements sufficiently to be prepared to respond, bring control, mitigate damage, and protect other interests. Risks come in all sizes and shapes. They can be foreseen. They can be planned for and mitigated. In fact, an entire line of management theory features risk management, including the multifaceted role of public relations. Thinking of risk, especially in terms of the pre-crisis phase, is a useful rubric for vigilance. What can happen? What is its likelihood? What would the effects be? Can it be mitigated—even prevented? How can it be explained and justified at least as a foundation for crisis response?

Heath (2004a) proposed several propositions relevant to effective pre-crisis preparation and management. For instance, as part of the communication planning:

> Organizations need to consider all of the publics that will want timely messages when a crisis occurs….Practitioners are wise to advocate the need to develop a crisis response plan, train to implement that plan, and practice implementing it….Each crisis constitutes a rhetorical problem that requires specific and tailored responses. (p. 34)

Efforts to manage crisis response begin with management. For that reason, consider these propositions:

> Planning should address the vulnerability of an organization as well as its strengths. Such is also the case for its ability to create and deploy messages and make other communication responses.
>
> Collaborative decision making that engages the organization's leadership to develop a comprehensive and useful plan can reduce the conflict that can harm implementation of the plan. (Heath, 2004a, p. 34)

One of the central themes in crisis is accountability, the willingness and ability of each organization to meet key stakeholder expectations on some matter that tests its willingness and ability to perform appropriately. Related to that theme is legitimacy. Crisis, by definition, alerts key publics to the possibility that the focal organization is unwilling or unable to meet the standard of accountability and is not doing what is necessary to achieve legitimacy.

Taking a perverse perspective on crisis, it is possible to frame a crisis not on what an organization did badly but what it did well but unwisely. Making huge profits, especially at a time when consumers are suffering, is a crisis of "good times" for the organization. If we think that crisis occurs because of what an organization does badly, what if its performance creates a crisis for other organizations? That argument can be made to help explain some of the organizations that caused the global financial meltdown of the past decade. Some prospered in ways that created a crisis for all; key managements took huge bonuses and even engaged in expensive executive retreats on taxpayer dollars.

How we define crisis determines whether we see its interconnection with issues, brand equity, and risk management. We might, for instance, focus on a crisis in the automobile industry if an assembly line worker is crushed, a vehicle model incurs dramatic failure in safety, or employees go on strike. But, would we see it as a crisis if the company engaged in inadequate risk management, whereby it committed too much capital to one model line (sports utility vehicles for instance) and too little to energy efficiency so that increases in fuel costs distorted the market and made the business plan obsolete? Which is more likely to cause the company to be in peril: a plant fire given dramatic attention on television or a misjudgment in the direction of the market?

Based on these logics, crisis communication is normative, as is its management—prevention and mitigation. The goal of management and communication is to prevent harm to others and to be accountable—and therefore legitimate participants in a community. Such endeavors have a proactive challenge to know, understand, and be able to identify and mitigate conditions that lead to crisis. The logic here is that crisis harms someone or something. We often prefer to think that the harm is to an organization's reputation, but it often is more than that. It harms some interest other than the organization, which in turn damages the organization's image. To repeat a now-trite phrase that summarizes the essence of this challenge: It's the harm, stupid! For this reason we explore another hypothesis: "Vigilant preparation can reduce the likelihood of a crisis and increase the responsiveness of the organization to demonstrate its ability to re-establish control over its operations" (Heath, 2004a, p. 35).

Recently, Coombs (2009a; see also Coombs & Holladay, 2010) offered additional perspective to our efforts to define crisis. He observed that at any one time, only a portion of the total crisis story is likely to reach the public, as people can only see a fraction of an iceberg:

> A crisis can be viewed as the perception of an event that threatens important expectancies of stakeholders and can impact the organization's performance. Crises are largely perceptual. If stakeholders believe there is a crisis, the organization is in a crisis unless it can successfully persuade stakeholders it is not. A crisis violates expectations; an organization has done something stakeholders feel is inappropriate. (p. 100)

As the focal point for assessing crisis response strategically, Coombs pointed out that such investigation and practice need to address (a) crisis management knowledge and (b) stakeholder reaction management.

Pre-crisis planning should consider multiple stakeholders and the crisis each may experience because of the action/inaction of an organization. Too often crisis management and communication look at the image/reputation and even issue impact of a crisis on an organization. Waymer and Heath (2007) reasoned that most crisis research focuses on a single organization and rarely on the larger set of entities that suffer in varying ways and to varying degrees of magnitude (the essential ingredients of a risk manifested). Sometimes the assumption is that all audiences witnessing, judging, and reacting to the focal organization are of one mind in such a way that a strategy can achieve universal impact with multiple publics.

Both errors in focus weaken the quality of research and development of best practices. Such is especially true if we think everyone is of the same mind and applies the same expectations of what can and should be said and done to put the crisis behind us. If the offending organization is a coal mining company, the focus might be on the company management, for instance, ignoring the crises experienced by the families involved, other community members, community including local government, vendors who supply mining equipment including safety equipment, and customers who depend on product flow. Surely insurance companies and regulatory bodies in various ways suffer crisis. Statements that might satisfy one set of victims (suffering their own crisis) and appease their concern might actually enflame a different audience, public, or other set of victims.

If we think of the crisis surrounding Hurricane Katrina, we quickly realize that no force could stop the storm. Crisis cannot, in that case, address prevention, but mitigation. Thus, for each crisis (thought of individually or by type), we need to be aware that we are interested in the various management, technical, and rhetorical problems at play—not only for the organization in the spotlight but also for others that are caught in the trajectory of this speeding bullet.

Exploring the power external organizations have during a crisis, Boys (2009) examined the rhetorical roles of two stakeholder groups in the Roman Catholic sex abuse case. She framed her case by arguing that the "U.S. Roman Catholic hierarchy lost exclusive jurisdiction over the situation" (p. 290). She analyzed the rhetorical contentions between the U.S. Roman Catholic Church, the Voice of the Faithful, and Survivors Network of those Abused by Priests. The defensive strategies used by the hierarchy were made more difficult because the voices of other groups framed the issue in terms of victims, who kept the issue salient, and who worked for larger interests than those only of church leadership.

Crisis as Process and Meaning

Both as process and meaning, crisis communication is rhetorical. It requires contextually relevant advocacy. Crisis discourse is propositional. As such, it entails the development (typically collaboratively rather than the work of one organization) of fact-based, evaluation-driven, and policy ripe conclusions. As facts, those offered must be such that they are sustained over time and under scrutiny by many other voices. They are likely to be stated in ways that invite counterstatements. Indeed, facts don't actually count as such as much as how they are framed and interpreted and whether they are accepted or rejected. One view of this discourse is that it is narrative (as in news stories and community resident "tales") of what happened: who did what (or did not do what), who was responsible, who were the victims, and how serious was the harm? What causal links are attributed because of this crisis narrative? As such, the discourse of crisis yields to the logics espoused by social construction and constitutive views of language and meaning.

Looking at the contributions of various approaches to crisis, Coombs (2009b) highlighted the contributions of rhetorical theory and cautioned that it might be too narrow—however rich its contributions. In addition, he stressed the role of more social scientific approaches to augment, refine, and enrich that approach. The key to such advances is that we should not be narrow in looking for the intellectual foundations of defining, clarifying, and adjudicating the roles of organizations in society.

Connecting Crisis, Risk, and Issues

For at least a decade, authors have toyed with the fact that crisis, risk, and issues are interdependent. At least since the toxic chemical release at a Union Carbide facility in Bhopal, India, risk management and communication has become a major discipline. Risks occur in various magnitudes, with varying degrees of predictability and as threats to identifiable parts of each society. Major theories have developed to understand, manage, and mitigate the impact of risk. It is a rationale for activism, government intervention, and corporate social responsibility.

As a risk manifests itself (such as Hurricane Katrina hitting New Orleans and other parts of the Gulf Coast), a crisis may occur. As such, the three phases of crisis preparation and response become relevant focal points for analysis and best practice. Risk simply is a probabilitistic assessment of what can go wrong and with what type of impact and magnitude. Crisis occurs when the risk manifests itself and people are harmed or worry that they are—and perhaps even wonder why they were not.

Issues can arise from risks. As such, an issue is a contestable matter of fact, value, policy, or identification. Known risks can be contested—magnitude, harm, occurrence, prevention, mitigation, and such—issues. An issue can become a crisis. One of the best examples is the issue over the safety/health hazards of tobacco use. As that was debated (the risk of health effects), it became a crisis for the tobacco industry and for public health authorities at state and federal levels.

A three-cornered connection between risk, issue, and crisis can have public policy implications and can arise from and lead to private sector threats and opportunities. A risk can create the opportunity for a product (a medication) or public policy (public health campaign). Toys, a vital part of seasonal giving and marketing, can pose risks (a crisis for parents and companies) and become a matter of public policy. This interconnection suggests the sorts of foundational themes that are becoming better known, understood, and tested to create a unique body of theory, research, and practice.

As such we have a growing body of insights with which to predict crisis and work to mitigate those which cannot be prevented. The case of NFL professional quarterback Michael Vick can help consolidate themes that will be explored in subsequent chapters.

Don't Be Vicked

The central objective of this brief case study is to enlarge and refine the analysis of crisis management and communication. It demonstrates that crisis analysis should not focus only on one organization, or in the case of Michael Vick, the person who is creating or suffering crisis. The design of crisis messages can too narrowly assume which ethical, responsible, and strategic responses are needed to inform key publics, achieve reputation restoration, and satisfy stakeholder expectations.

Narrow analysis for pre-crisis management and communication planning is likely to suffer managerial bias (Waymer & Heath, 2007), which ignores or marginalizes other voices victimized by the crisis. If such outcome results, the planning and execution of the crisis response become distorted, less insightful, and finally less encompassing of the interests that need to be addressed in crisis resolution and renewal.

In the case of superstar quarterback Michael Vick, not only was his career and brand equity damaged by his association with dog fighting, but he also created a crisis for the NFL. Given the three narratives that emerged from the crisis (Vick's, the dogs' through People for the Ethical Treatment of Animals, or PETA, and the NFL's), Vick found himself without a strong narrative that he could use to protect himself and his brand equity. The key theme in what follows is that the NFL needs stronger issues management training to provide its athletes with a reduced likelihood they will encounter a crisis. The League also needs a solid pre-crisis prevention and response plan for the eventuality that it cannot create and sustain a crisis response narrative that protects it and its key athletes.

Multi-million Dollar NFL Brand

Heath and Palenchar (2009) explained that a brand is "important for individual or corporate personalities because it represents what others outside the organization view or perceive the person or company to be" (p. 355). Part of athletic performance—connecting a player's brand equity and that of the NFL—is the character exhibited by star athletes. They are touted as role models—as evincing high standards that should be emulated. Perhaps one of the most important aspects of Vick's situation is that "the brand equity as the reputation of an organization is its authority to speak on some matter and the power resources it can bring to bear on a market purchase, share value, or a governmental policy" (p. 358).

Heath and Palenchar (2009) underscored that brand equity is "a consideration of both market value and book value…an idiosyncratic blend of customer preferences and the related associations and relationships with the product, service or organization" (p. 358). Relevant to brand equity, Benoit (1997) stressed, *image* is the perception of a person (or group or organization) held by the audience, shaped by the words and actions of that person, as well as by the discourse and behavior of other relevant actors. Higgins and Snyder (1989) explained that people have "a basic need to maintain positive images" (p. 74). Schlenker (1980) recognized that predicaments can "damage" one's identity or image, "adversely affecting relationships with the audience" (p. 131).

Vick utilized different image restoration techniques as part of his post-crisis response strategy (Benoit, 1997; Coombs & Holladay, 2008; Hearit, 2006). Before being challenged by an emergent voice, his script utilized standard—albeit poor—forms of crisis response strategies like apologia, denial, evasion of responsibility, and others. Since the dogs were mute victims of the crisis, those crisis response strategies were a biased yet momentarily effective way to handle the situation. However, the voice PETA provided for the victims, the dogs, addressed dog fighting in such a way that it not only challenged Vick's brand equity but also his sponsors' and the NFL's brand equity and image (Sultan, 2009).

Coombs and Holladay (2008) suggested that often "the value of an apology is established by comparing how people react to an apology and to other less victim-centered/accommodative responses. This 'unfair' comparison results in apology appearing to be 'the' best alternative" (p. 252). Similarly, image restoration strategies suggested by Benoit (1997) may be ineffective crisis responses, particularly when emergent voices arise.

Vick-tim or Villain?

Prior to the dog fighting incident that eventually sent him to prison, Vick was not immune to scandal, though much of it was dismissed, perhaps due to the lack of an emergent third voice (victim) and through implementation of a well-executed routine script on behalf of his public relations team.

Vick reached a settlement in 2006 with a woman who charged in a lawsuit that he had knowingly given her herpes. That same season, Vick was fined $10,000 by the NFL and agreed to donate another $10,000 to charity for making an obscene gesture toward fans while leaving the field after a disappointing loss at the Georgia Dome in late November.

In January 2007, authorities said that a water bottle surrendered by Vick at a security checkpoint at Miami International Airport had a marijuana-like substance in a secret compartment he said was used to store jewelry. Later it was announced that laboratory tests showed no evidence of drugs and no charges would be filed.

The key event, for our analysis, occurred on April 27, 2007. During a drug raid, police found a dog training complex in the woods behind Vick's home in Virginia with 66 dogs on the property. Court documents alleged that the dog fighting operation, Bad Newz Kennels, had operated on his property for five years. Shortly thereafter, Vick and three of his associates, Purnell Peace, Quanis Phillips, and Tony Taylor were indicted by a federal grand jury. All four were charged with competitive dog fighting, an activity they conducted across state lines. The 19-page indictment alleged Vick was highly involved in the operation, stating that he attended fights and paid off bets when his dogs lost. It said he also was involved in the executions of dogs that did not perform well. Dog fighting is a felony in Virginia and 47 other states. Transporting dogs over state lines for dog fighting is a federal crime. If convicted, Vick would face up to six years in prison and $350,000 in fines.

Initially, Vick denied everything, which at the time, depending on if he was telling the truth or not, was either a good, standard, response strategy or a horrible one. One of the keys to crisis, however, is the emergent voices that surface over time and whether they confirm or disconfirm narratives that various other parties are seeking to advance.

Four days later, Taylor, Peace, and Phillips entered guilty pleas and agreed to testify against Vick. They said they traveled with Vick to dog fights and described, in detail, the execution by hanging and drowning of approximately eight dogs that did not perform well. The narrative crisis response employed by Vick began to suffer counter-narrative challenges as one of the characters, as defined by him, took on a voice that collided with the portrayal presented by Vick. As Coombs (1999) indicated, the timing and completeness of information given out are also essential to maintaining (in this case, Vick's) reputation and credibility throughout the crisis situation.

Coombs suggested that "if the crisis team does not supply the initial crisis information to the media, some other group will" (p. 115).

In this case, by the time Vick was indicted, PETA had already saturated the media with an appeal voiced on behalf of the victims—the dogs, which would otherwise have remained mute. On their website, PETA called upon supporters to protest Vick's behavior with severe punishments and called on "NFL Commissioner Roger Goodell to immediately suspend Michael Vick in light of this indictment" (The PETA Files, 2007).

By launching nationwide media campaigns with graphic images portraying dog fighting, PETA's mission to provide the canine victims a voice chipped away at Vick's denial and then apologia. The emergent voice of the victims allowed for dialogue and accountability to emerge. PETA's plan aimed at publicly pressuring the major parties involved in the case—the NFL league and its commissioner, Vick's supporters, sponsors, and even Vick himself—to be held accountable for the crimes committed, even if only alleged. By giving the victims a voice to be heard publicly and also appealing to the American people, of whom almost 40 percent own at least one dog, PETA held Vick responsible for his actions.

As these other voices emerged, at stake was not only Vick's freedom but also his career with the NFL, his endorsement deals with various companies, his credibility with fans, and, ultimately, his present and future brand equity. When the evidence and reputational concerns mounted, Vick ultimately signed a plea agreement and statement of facts admitting to conspiracy in a dog fighting ring and helping kill pit bulls. He denied betting on the fights—only bankrolling them—but by that time, the damage had already been done.

Lesson Learned

Although the analysis above features failed crisis response, the case study offers significant pre-crisis lessons to sports celebrities like Vick and organizations like the NFL—and other individuals and organizations. It demonstrates the deficiency of Benoit's (1997) image restoration techniques as being narrow, especially if an emergent voice generates counter-narratives. In such circumstances, the voice of the offender is either likely to fail or be less successful if it does not appreciate interests beyond its own and in ways that address the fullness of the crisis. Thus, and it is often the culture of the individual or organization that precipitates its dysfunctional planning and response, the case indicts the forgiving culture that surrounds star athletes: "Boys will be boys." Whether it's banks too big to fail, industries too vital to fail, or athletes too good to punish, pre-crisis analysis and planning must examine the limiting premises that otherwise seem to support responses such as denial or apology.

In the case of professional sports, if management is unprepared to engage in and demand of athletes and other voices a more embracing pre-crisis awareness and their role in it, they are likely to suffer from miscalculation. Thus, understanding pre-crisis conditions and either avoiding them or learning what needs to be planned in order to achieve an effective and responsible response is vital to success. The fact that Vick was left alone to deal with his financial and legal downfall prompts the recommendation that the NFL, its players, and other stakeholders and player organizations prepare more adequately for crisis situations—primarily by prevention. As the face and image of the industry, NFL athletes require much more attention to develop and protect not only their own brand equity but also that co-branded with the NFL. And athletes, in this sense, should be disabused of the notion that they are above reproach. In fact, they may be less so than others because they are touted, their brand equity, as role models.

Organizations like the NFLPA aid players in different ways, including retirement needs and financial planning. They can help make a transition into a broadcast career, business management and entrepreneurial programs, coaching internships, career transitions, and continuing education workshops. However, nowhere does the NFLPA seem to provide transitional support between collegiate and professional football, psychological and emotional support, workshops on communication strategies, social networking, legal and illegal behavior, and more. In addition to additional player development programs, the NFL industry must consider and formulate positive relationships with advocacy groups that address the potential issues put forth by their players (anti-drunk driving, domestic violence, and anti-gun groups, for example). The NFL's proactive relationships with these groups can help minimize risk and the mismanagement of a potential crisis. Had Vick worked with advocacy groups and been honest with the NFL and the public initially, he may have received a lesser sentence. Pre-crisis planning should also include monitoring the internal and external activity of the NFL, which helps define its mission, goals, and objectives and prevents issues from developing into crises. The explicit connection illustrates how monitoring (both risks and issues) seeks trends among stakeholder concerns, interests, and attitudes that affect brand equity of an organization by reevaluating its "relevance and believability" (Heath & Palenchar, 2009, p. 361).

The interconnectedness of these three organizational functions—risk, crisis, and issues management—is essential to a company's survival and maintenance. Heath and Palenchar (2009) explained that issues management and, essentially pre-crisis training, could help organizations plan and manage themselves by "making strategic adaptations needed to achieve harmony and foster mutual interests within the communities where they operate" (p. 9). Not only does this aid an organization's survival, but it also provides an opportunity for growth and to "lessen the threats that public policy trends and marketplace claims pose to their strategic business planning" (p. 9).

Michael Vick's brand equity was carefully honed by his performance and communication efforts, including those of the NFL and the companies whose products and services he had endorsed. Since completing his sentence and reentering the NFL, Vick has proven that an effective post-crisis plan can offer a second chance to an organization or individual who has suffered a devastating crisis. President Obama recently called positive attention to this rehabilitation. The NFL can learn from Vick's rise and disastrous fall from the public's grace. Moreover, organizations outside of the sports industry can utilize the lessons and points made in regard to emergent voices in crisis and pre-crisis management.

Concluding Thoughts and Final Challenges

Addressing crisis narrowly as communication challenge and only focusing on reputation restoration ignores the magnitude of the challenge. If the "crisis" truly is one, then survival can be on the line—not merely a tarnished reputation but even the opportunity for renewal (Ulmer et al., 2009). Also, taking a view of crisis communication that is linear and focusing on what the crisis manager says can miss or avoid the complexity of crisis. Gilpin and Murphy (2008) "urge a paradigm shift for crisis management in which uncertainty, adaptiveness, and improvisation replace certainty, goal orientation, and control" (p. 177). Such logics are a foundation of participative management, which is often seen as a philosophy best able to shape internal organizational culture. Crisis communication, as other aspects of public relations and strategic communication, serves society best when it helps it to function more fully.

What we have outlined in this chapter stresses the importance of reflective vigilance on the part of management to spot potential crises. Once that is done on an ongoing basis, planning and implementation need to be put in place to prevent, mitigate, and respond as or when crisis occurs. Communication is part of that process as collaboration inside and outside the organization occurs. For instance, as a high-risk facility, such as a petrochemical plant, works to spot and mitigate crises, it needs to communicate with near neighbors as well as emergency response and government officials. It is not just an organizational challenge but also a community challenge. The organization should take the lead and work with the community.

Empowerment is required so that requisite variety can operate to increase the organization's vigilance and response. All of the crisis response options typically discussed in the literature—practitioner and academic—can have greater crisis response efficacy if careful planning and management provide a foundation. It is easier to deal with a situation where high safety standards, for instance, were developed and implemented rather than when known violation was typical organizational protocol. Reputation, or brand equity, is best guarded by effective pre-crisis preparation. So, too is the sort of issues debate that can occur in the aftermath of something characterized as crisis.

The chapters that follow offer insights, theories, research conclusions, and refined best practices. Such analysis can help managers to prevent and mitigate crisis. They can lead to more ethical and reflective responses that move us closer to stakeholder expectations of what is necessary for organizations to deserve the right to operate in the interest of others, as well as its own interest.

References

Benoit, W. L. (1997). Hugh Grant's image restoration discourse: An actor apologizes. *Communication Quarterly, 45*, 251–267.

Boys, S. (2009). Inter-organizational crisis communication: Exploring source and stakeholder communication in the Roman Catholic clergy sex abuse case. In R. L. Heath, E. L. Toth, & D. Waymer (Eds.), *Rhetorical and critical approaches to public relations* (pp. 290–300). New York: Routledge.

Coombs, W. T. (1999). *Ongoing crisis communication planning, managing and responding.* Thousand Oaks, CA: Sage.

———. (2009a). Conceptualizing crisis communication. In R. L. Heath & H. D. O'Hair (Eds.), *Handbook of crisis and risk communication* (pp. 100–119). New York: Routledge.

———. (2009b). Crisis, crisis communication, reputation, and rhetoric. In R. L. Heath, E. L. Toth, and D. Waymer (Eds.), *Rhetorical and critical approaches to public relations* (pp. 237–252). New York: Routledge.

Coombs, W. T, & Holladay, S. (2008). Comparing apology to equivalent crisis response strategies: Clarifying apology's role and value in crisis communication. *Public Relations Review*, *34*, 252–257.

Coombs, W. T., & Holladay, S. J. (Eds.) (2010). *The handbook of crisis communication.* Malden, MA: Wiley-Blackwell.

Fisher, W. R. (1987). *Human communication as narration: Toward a philosophy of reason, value, and action.* Columbia, SC: University of South Carolina Press.

Gilpin, D. R., & Murphy, P. J. (2008). *Crisis management in a complex world.* Oxford: Oxford University Press.

Hearit, K. M. (2006). *Crisis management by apology: Corporate response to allegations of wrongdoing.* Mahwah, NJ: Lawrence Erlbaum.

Heath, R. L. (2004a). Crisis preparation: Planning for the inevitable. In D. P. Millar & R. L. Heath (Eds.), *Responding to crisis: A rhetorical approach to crisis communication* (pp. 33–35). Mahwah, NJ: Lawrence Erlbaum.

———. (2004b). Telling a story: A narrative approach to communication during a crisis. In D. P. Millar & R. L. Heath (Eds.), *Responding to crisis: A rhetorical approach to crisis communication* (pp. 167–188). Mahwah, NJ: Lawrence Erlbaum.

Heath, R. L., & Millar, D. P. (2004). A rhetorical approach to crisis communication: Management, communication processes, and strategic responses. In D. P. Millar & R. L. Heath (Eds.), *Responding to crisis: A rhetorical approach to crisis communication* (pp. 1–17). Mahwah, NJ: Lawrence Erlbaum.

Heath, R. L., & Palenchar, M. J. (2009). *Strategic issues management: Organizations and public policy challenges* (2nd ed.). Thousand Oaks, CA: Sage.

Higgins, R. L., & Snyder, C. R. (1989). The business of excuses. In R. A. Giacalone & P. Rosenfeld (Eds.), *Impression management in the organization* (pp. 73–85). Hillsdale, NJ: Lawrence Erlbaum.

Jaques, A. P. (2009). Issue management as a post-crisis discipline: Identifying and responding to issues beyond the crisis. *Journal of Public Affairs*, *9*, 35–44.

Millar, D. P., & Heath, R. L. (Eds.) (2004). *Responding to crisis: A rhetorical approach to crisis communication.* Mahwah, NJ: Lawrence Erlbaum.

Schlenker, B. R. (1980). *Impression management: The self-concept, social identity, and interpersonal relations.* Monterey, CA: Brooks/Cole.

Schwartz, K. A. (2010). Red Cross crisis communication in the wake of September 11, 2001. In R. L. Heath (Ed.), *Sage handbook of public relations* (pp. 501–508). Thousand Oaks, CA: Sage.

Sultan, J. (2009). (Master's thesis). *Narrative analysis of crisis: "Beware of dog" or be blitzed by the Vick-timized voice.* University of Houston, Houston, TX.

The PETA Files. (2007, July 18). Vick's indictment: What you can do [Blog]. Entry posted to http://blog.peta.org/archives/2007/07/vicks_indictmen.php

Ulmer, R. R., Sellnow, T. L., & Seeger, M. W. (2009). Post-crisis communication and renewal. In R. L. Heath & H. D. O'Hair (Eds.), *Handbook of crisis and risk communication* (pp. 304–324). New York: Routledge.

Waymer, D., & Heath, R. L. (2007). Emergent agents: The forgotten publics in crisis communication and issues management research. *Journal of Applied Communication Research*, *35*, 88–108.

Chapter 7

Pre-crisis Threat Assessment: A Cognitive Appraisal Approach

Yan Jin, Augustine Pang & Glen T. Cameron

These authors significantly expand our understanding of threats in crisis situations and further distinguish them from risks, fear, and conflict. They offer a detailed examination of how threat assessment begins with an understanding of the antecedents of the crisis situation, which helps foster understanding of how the threat evolves during the crisis situation. Diagnosing the threat stems from evaluation of the threat type, level, and duration. The authors propose that threat consequences be considered from cognitive, affective, and conative perspectives.

Introduction

Threat has been commonly used to describe the state of seizure a nation, organization, or individual experiences before and during a crisis, which has been explicated in crisis communication research conceptually (Jin, Pang, & Cameron, 2005; Rosenthal & Kouzmin, 1997) and empirically (Hwang & Cameron, 2008a, 2008b; Jin, 2010; Jin & Cameron, 2007; Pang, Jin, & Cameron, 2006). Threat, which is often used interchangeably with *risk*, *fear*, and *conflict*, is both the cause and the effect of crisis and is particularly important in issues monitoring and pre-crisis preparation. This chapter examines a pre-crisis threat assessment model by introducing a cognitive appraisal approach to evaluate the situational demands and allocated organizational resources to enhance organizational preparedness and to address the pre-crisis issues more effectively. This understanding of the faces and fabric of threat before and during a crisis has enormous implications for theory development in the understanding and diagnosing of threats in crisis, which, in turn, affects how crisis can be communicated, managed, and survived by organizations.

With an increase in terrorism activities on U.S. soil after 9/11, the Department of Homeland Security, in a bid to alert U.S. citizens on the status, nature, and level of threat, has devised a color-coded system: green for low risk, blue for general risk, yellow for significant risk, amber for high risk, and red for severe risk (Department of Homeland Security, 2004).

Yet, others have regarded threat as a less ominous, amorphous, but nonetheless, severe nature. An accusation can be regarded as a threat to one's image (Benoit & Pang, 2008); so can structural malfunctions be a threat to one's plant. Indeed *threat*, in all its variations as a noun, verb, or adjective, has been used to describe the state of seizure a nation, organization, or individual is in during a crisis. A threat causes and leads an organization into crisis. Crisis, as defined by Kathleen Fearn-Banks (2007), is "a major occurrence with a potentially negative outcome affecting" (p. 8) an organization. It threatens the "existence of the organization" (p. 8), which, in turn, disrupts and interferes with the organization's ability to function, its ability to deliver its products and services, financial stability, its relationship with its publics, and its image and reputation. Threat thus becomes both the cause and the effect of a crisis.

Rosenthal and Kouzmin (1997) argued that threat is closely associated with crisis and, by extension, crisis response in three ways. First, the threat determines the kind of crisis. A threat to a building's sewerage system, for instance, more often than not limits the problem to a crisis of structural problems. It is argued that some threats have the propensity to escalate and elevate the crisis to unimaginable heights disproportionate to the basic threat. Second, the authors argued that the domain of threat can be viewed in geographical terms (i.e., within a certain organization, locale, or region), and the crisis can be contained within that domain. Some events, the authors conceded, did have spillover effects. Third, the origins of threats could either be "endogenous or exogenous to the system affected" (p. 230). Some crises could be contained from within and others from without.

In times of highly threatening situations, small groups and larger organizations—even nations—have been found to "close ranks" and accept strong, dominant leadership, concentrated to a high degree in the hands of one man (Mulder & Stemerding, 1963; Volkan, 1997). Three consequences of threat are predicted: a need to lean on strong leadership, a derived tendency to counteract other striving group members, and a need to associate with others. In order to fully understand the concept of threat, many questions are yet to be answered, such as: What is considered a threat? How can threat be measured? In a crisis, what are the options available to the organization to counter the threat?

Threats in Crisis

Crisis literature has made vague references to its existence and response to it without delving fully into it. Ray (1999), for instance, argued that threat could be controlled with effective crisis management. Richards (2004) equated threats as terrorism based and called for greater understanding on how public relations can be utilized to counter the threat of terrorism. Wrigley, Salmon, and Park (2003) argued that large corporations should make room for crisis management planning to take into account the threat of bio-terrorism. Coombs (1998) argued that an appropriate organizational response to threats in a crisis should take into account two key variables: if the organization has strong or weak control over the crisis and if the organization is perceived to be strongly or mildly responsible for the crisis.

Benoit and Pang (2008) argued that threats affect the image of the organization. The image or reputation of the organization is important as it determines if the organization is trustworthy in the eyes of stakeholders. The organization's image also influences the extent to which government authorities regulate its actions. Thus, threat is associated with loss of image and reputation, and threats to the image of the organization are "ubiquitous" (p. 244), the authors argued.

As far as theories go, threat was alluded to in Grunig's excellence theory. The theory offers four models where public relations can be carried out: the press agentry, public information, two-way asymmetrical, and two-way symmetrical models (Grunig & Grunig, 1992). Based on the two-way symmetrical model in the theory, threat could be "smoothed over," if you will, if both the organization and its publics can engage in meaningful dialogue to achieve a win-win solution.

The contingency theory of strategic conflict management, which argued that two-way symmetrical communication may not always be possible and takes an alternative perspective of how communication ought to take place—through a continuum rather than by positing ideal models of excellence (Pang, 2008)—provided a comparatively deeper insight into the concept of threat. Threat is a "potentially negative situation involving publics…the greater threat a public presents to an organization, the faster the organization will respond to the public and the more accommodating the organization will be to the public" (Cancel, Mitrook, & Cameron, 1999, p. 184). The contingency theory distinguished two kinds of threats: external and internal threats facing the organization (see Appendix). Internal threats include (1) economic loss or gain from implementing various stances; (2) marring of employees' or stockholders' perception of the company; and (3) marring of the personal reputations of the company decision-makers. External threats include (1) litigation; (2) government regulation; (3) potentially damaging publicity; (4) scarring of company's reputation in business community and among the general public; and (5) legitimizing activists' claims. Subsequent studies found that the threat of damaging publicity supplanted all the other threats, followed by threat of government regulation and litigation (Pang, Jin, & Cameron, 2010). Beyond this, the theory has not fully explained how these threats can be so deleterious in a crisis.

While this has provided a map to understand how threats work, research on the *antecedents* of a crisis and how a perceived threat might dynamically evolve and influence the crisis situation has been scarce. Grunig and Huang (2000) attempted to do so by developing a theoretical model of the antecedents of relationships, the strategies public relations practitioners could use to develop and maintain relationships, and the outcomes of relationships. For each stage in the model, the authors suggest indicators that can be used to identify antecedents, to monitor ongoing public relations processes used to maintain relationships, and to measure and evaluate the long-term effects of public relations programs. The antecedent factors in a relationship model can be used to explain the emotion process in public relations by investigating the situational conditions of a public relations encounter and the characteristics of both the public and the organization in the encounter as well as how these two parties interact for a solution. This model, however, does not provide a holistic view of the full nature of threats.

Unveiling the Fabric of Threat in Crisis

Besides the conundrum and confusion over the nature and effects of threat, we also found that in examining a crisis, *threat*, *risk*, *fear*, and *conflict* have been used interchangeably without sufficient distinction or operationalization.

To understand how threat is featured in organizational crisis, a clearer explication of the concept is critical. First, we propose to delve into the study and explication of threat in crisis by integrating ideas not just from the works of communication and public relations but also psychology and conflict management literature. Coombs (2001) argued that this fusion will form a "rich tapestry of literature" that would provide a "broader perspective" (p. 99) for people to understand the nature of crisis. Second, we propose a threat appraisal model. Third, we illustrate the dimensionality of a threat in crisis situations. Finally, we propose a series of testable propositions with practical implications for researchers and practitioners.

In doing so, we believe this would contribute to the growing understanding of threat. A rooted conceptual understanding of threat in crisis will advance the theory development in crisis communication, which leads to better understanding and diagnosing threats in crisis. This, in turn, affects how crisis can be managed with optimal effectiveness.

Conceptualizing Threats in Crisis

Threats, Fear, and Risks

Threat appeals have been widely used by communication practitioners in designing effective threat messages and communication campaigns (Strong, Anderson, & Dubas, 1993). In their definition, a threat is a warning to recognize danger and accept the recommendation to avoid the negative consequences. They advocated that it is important to differentiate between threat as a stimulus and the recipient's cognitive and emotional responses to that stimulus. Threat appeals sometimes are used interchangeably as fear appeal or fear arousal. However, based on the distinction between threat as a stimulus and the outcome of threat, we can see that fear is only an outcome of perceived threat. Different levels of threat may lead to different emotional responses.

Risk is another concept sometimes used as a synonym for danger or threat. However, they are different concepts. Risk is usually defined statistically to provide a relative measure of safety, which should be more precisely regarded as a synonym for expected loss as defined in decision theory (Oppe, 1988). Unlike threat, risk is the "expected loss" of an alternative to be chosen, which concentrates on the process of decision-making and not solely on the result. Risk is not loss but expected loss, which strongly connects with the decision-maker's behavior. By collecting and processing information about the situation, one can reduce the uncertainty about the expected outcome of each possible action (Oppe, 1988). From the public relations' point of view, risk is the expected loss if a particular strategy is chosen, given the information available. The real loss is uncertain and is only known afterward. Thus, a hallmark of risk communication, argued Wilcox and Cameron (2009), is to reduce the danger and threat posed. As Oppe (1988) mentioned, the relation between the information and the choice of an action is called a strategy, and a strategy will be chosen such that, given the information, the expected loss is minimal. In this sense, the uncertainty attached to risk can be a component of threats.

Threats in Power Relations

Power is essentially a structural concept, referring to certain aspects of the functional arrangements of any social system, and at the same time it deals with the motivations of individuals (Cohen, 1959). The exercise of power requires some acceptance by those who are part of the social system, and because reward and punishment are inherently involved, it has repercussions upon the adjustment of and reactions in the power relations between parties. With regard to threats, studies have been done to determine the amount of threat experienced by a low-power person in one's relations with someone of higher power (Cohen, 1959). It was found that a person's reaction to being under the power of another depends to a considerable degree on one's view of the self (i.e., self-esteem).

Cohen defined *power* as "the ability of one party of a relationship to determine whether or not the other party is carried toward his goals or away from them, over and above the second party's own efforts" (p. 36). In public relations, power can be taken as the ability of one party of a relationship to influence the behavior of the other party. In many situations, one party's power consists of the ability to determine whether or not another party reaches its goal, which amounts to the ability to control the means by which the other approaches the goal.

As Cohen (1959) further theorized, threat may have many sources. One possibility is to put the perception of the adequacy of oneself to deal with the situation and to satisfy one's needs on a continuum. In this sense, threat can be defined as "the state of the individual in which he feels inadequate to deal with a given situation and to satisfy his needs" (p. 36). Combining the definition of power and threat, it seems that anyone under the power of another would experience threat. However, the possession of power consists of the potential to withhold or to permit the gratification of a need. Therefore, the degree of threat one may potentially experience in a social setting is related to the degree of power which is exercised over him.

Closely related to the concept of power, *structure* refers to the degree to which a social situation provides the individual with clear and accessible cues so that he may behave in a goal-directed and need-satisfying manner.

Indeed, such a structure is deemed to be essential, as evident in public relations literature, to address the imbalance caused by threat. Since communication is conceived as an activity to improve one's image (Benoit & Pang, 2008), the individual must, when given the chance, work within socially accepted norms to regain one's reputation when he or she is threatened, either by accusations or an offensive act. This is because accusations threaten one's face, image, and reputation, and when one's face, image, and reputation are rendered in question, it limits one's ability to influence the other (Benoit & Pang, 2008). In understanding the impact of threat to one's reputation, the key question, argued Benoit and Pang (2008) then, is to immediately address the offensive act that is regarded by one's audiences. This is done when one offers explanations, defenses, justifications, rationalizations, apologies, and excuses for one's behavior. To do so, Benoit and his colleagues have developed the theory of image repair discourse to repair one's image (see Benoit, 1995, 2004; Benoit & Pang, 2008).

Threats and Conflicts

How to deal with threats and threatening situations has been an intriguing and important facet of conflict management and crisis communication research. Maslow (1943) began to study conflict, frustration, and the theory of threat in his early work, in which the outcome of threat was categorized into cognitive outcomes and affective outcomes manifested by fear and frustration. The understanding of these key concepts was highlighted as very important, especially in terms of coping strategies in a conflict situation.

According to Maslow (1943), there are two kinds of conflict: threatening and non-threatening. In explaining the occurrence of threats, Maslow took the view of psychopathology: The most unclear aspects of threat are the direct deprivation of (or thwarting) or danger to the basic needs. Tightened by his motivation theories, Maslow advocated that we must ultimately define a threat situation in terms of the individual organism (e.g., organization) facing its particular problem. It was also emphasized that the feeling of threat in itself should be considered a dynamic stimulation to other reactions. By the same token, to better understand threats in a conflict or crisis situation, we have to examine not only their antecedents but also their nature or dimensionality as well as the consequence of how an organization will react to deal with the situation.

Analyzing Threats in a Crisis: A Cognitive Appraisal Model

Studies argued that the perception of threat is not strictly a function of an environmental stimulus itself, but it involves an interpretation of the stimulus (Carver, 1977), which corresponds to Jin and Cameron's appraisal model of emotions in public relations (2003).

According to Jin and Cameron (2003), appraisal refers to the evaluation of the significance of what is going on in the public-organization relationship for each party's benefit, which leads to different patterns of action tendency and coping process. Coping refers to how the public and the organization try to manage the relational encounter so as to alter a troubled relationship or sustain a desirable one. It consists of cognitive and behavioral efforts to manage specific external or internal demands (and conflicts between them) that are appraised as exceeding the resources of the public or the organization. There are two types of coping: problem-focused coping, which changes the actual relationship between the public and the organization via actual measures and steps, and emotion or cognitive-focused coping, which changes only the way that the relationship is interpreted by practitioners.

Threats as a Result of Skewed Relationship Between Demands and Resources

The threat appraisal has been proposed by Blascovich and Mendes (2000) as a cognitive appraisal process consisting of *primary* appraisals (of situational demands) and *secondary* appraisals (of the individual's resources). Threat appraisal consists of the interplay between demand and resource appraisals. More specifically, demand appraisals involve the perception or assessment of danger, uncertainty (situational versus task uncertainty), and required effort inherent in the situation. On one hand, perceptions of high demand on any one of these dimensions may trigger high overall demand appraisals. On the other hand, resource appraisals involve the perception or assessment of knowledge and skills relevant to situational performance.

Psychological research demonstrated that affective cues influence the experience of threat. Blascovich and Mendes (2000) attempted to integrate purely affective and cognitive processes into their biopsychosocial model of threat. By definition, threat represents person/situation-evoked motivational states that include affective, cognitive, and physiological components. According to Blascovivch, Mendes, Hunter, and Salomon (1999), as motivational states, threat is related to approach-avoidance or appetitive-aversive states. On the affective level, this involves positive and negative feelings and emotions, and on the cognitive level, it involves attention and appraisal.

Thus, threats occur when the decision-maker in the organization experiences insufficient resources to meet situational demands. Threat process begins in a situation in which a decision-maker expects to perform. In that sense, the person must perceive the situation as goal-relevant and evaluative and believe that adequate performance is necessary to his or her continued well-being or growth as well as that he or she will undergo evaluation in this situation either by others or by oneself on some important self-relevant domain. In a crisis, Pang, Jin, and Cameron (2006) presented a model to understand the elements that compose a given threat: For a threat to occur, there has to be some insufficiency of resources to meet situational demands. Thus, resources need to be specified as resources that can be allocated at the moment; otherwise the crisis might be an unsolvable disaster. The threat appraisal process in a crisis is initially composed of a primary appraisal (situational demands) including *Danger, Uncertainty* (lack of prediction and control make it difficult for meeting adequately); *Required Effort*; and a secondary appraisal (resources), which includes *Knowledge and Skill*, *Time, Finance*, and the *Support from the Dominant Coalition*. Noticeably, the perceived uncertainty may influence the way a practitioner or an organization assesses the required effort and danger in the crisis situation.

Dimensionality of Threats

Lanzetta, Haefner, Langham, and Axelrod (1954) studied the influence of a threat on group behavior. They operationalized the variation in threat along several dimensions: nature and/or intensity, locus in time, locus in "psychological" space, and target. The latter two are peculiar to a concern with group processes and have no counterpart if the focus was the individual acting alone. In their experiments, the authors manipulated the threat condition as external and internal. Similarly, Carver (1977) mentioned threats may come from external agents, or they may be self-imposed.

Management literature discussed three types of threats (Strong, Anderson, & Dubas, 1993): (1) physical/social threats; (2) immediate/delayed threats; and (3) second/third party threats. From a public relations perspective, we shall focus on the social threats in the communication management process with different publics. Secondly, immediate/delayed threats can be further developed into durations of threats and the perceived levels or intensity of threats. Finally, it would be better to understand the target of threats in terms of whether they are external or internal. Therefore, we propose Pang, Jin, and Cameron's (2006) three dimensions for each given threat.

Dimension 1: Types of Threats. Borrowing insights from the contingency theory, the two types of threats identified are used here (Pang, Jin, & Cameron, 2010): external threats and internal threats. External threats include litigation, government regulation, potentially damaging publicity, scarring of the organization's reputation in community, and legitimizing of activists' claims. Internal threats include economic loss or gain from implementing various stances, marring of employees' or stockholders' perception of the company, and marring of the personal reputations of the company decision makers (image in employees' perceptions and general public's perception).

Dimension 2: Levels of Threats. Levels of threat have been operationalized as low, medium, and high in previous studies (Corneille, Yzerbyt, Rogier, & Buidin, 2001).

Dimension 3: Duration of Threats. This dimension refers to the longitudinal facet of a threat, which means whether the threat is perceived as a short-term one or long-term one.

Consequences of Threats

Based on the above appraisal mechanism of a threat as well as the three proposed dimensions of a threat, we now move on to the possible outcomes or consequences of threats. According to Strong, Anderson, and Dubas (1993), threat stimuli lead to the following occurrences: perceived threat, cognitive and emotional processing, beliefs and feelings, attitude toward the threat, and intentions. Behavior is the ultimate outcome variable. A threat stimulus appears, and one may perceive the threat represented by the stimulus. The perceived threat leads to cognitive and subjective processing, which affects specific salient beliefs (e.g., perceived severity) and relevant emotions (e.g., fear). These beliefs and emotions combine to form the consumer's attitude toward the threat, which then leads to intention and behavior.

Lanzetta, Haefner, Langham, and Axelrod (1954) further mentioned that threat appears to result in a reduction of forceful, assertive, aggressive, interpersonal, and task-directed behavior; in a greater concern with group acceptance, it can result in poorer group effectiveness and greater variability in effectiveness. Strong, Anderson, and Dubas (1993) argued that perceived threat leads to two types of processing: cognitive and emotional, which demonstrated salient beliefs about the threat and the emotional arousal. These two processes may interact with each other. Cognition provides direction and continuity, while emotions provide intensity and urgency in the threatening situation. Carver (1977) further proposed that one's attempts to deal with threats depend on the nature of the threat and the freedom on which it is imposed. Possible outcomes would be changing one's attitude in a direction opposite to that advocated by a threatening communication, expression of hostility toward the threatening agent, derogation of non-threatened options, and increased attraction toward threatened options.

For a crisis situation, we propose the following consequences. First, at the cognitive level, the practitioner has to perceive the severity of the crisis and weigh the demands in deciding how to allocate resources to manage it. Second, at the affective level, the practitioner will be exposed to an arousal of threat about the situation (i.e., how negative he or she feels about the crisis, what would be the aroused level or the intensity of being threatened, and how important this threat is to decision-making at the moment). Third, at the conative level, it impacts on what stances the practitioner will take in moving the organization along the continuum of advocacy versus accommodation as propounded by the contingency theory (Pang, Jin, & Cameron, 2010), as well as what kind of crisis communication or management strategies he or she is going to apply in order to deal with the crisis for the organization.

Given the upheavals practitioners are likely to face, we propose the following propositions.

Proposition 1: Cognition—Perceived Demands and Sources. Cognitive processing of threats is proposed to have the following salient beliefs (Strong, Anderson, & Dubas, 1993): severity, vulnerability, efficacy of dealing with the threats, alternative coping options, etc. These determine the outcome of attitudes toward the threats. More specifically, the more severe the threat, the more motivated one is for avoidance. Severity beliefs are the perceived costs of the threatening situation. Vulnerability is the perceived probability of the threat occurring; higher perceived vulnerability increases the likelihood of compliance.

We propose that perceived severity and vulnerability are components of perceived demands of a threat, while the organization's decision-makers' self-efficacy is one indicator of the sources to deal with a given threat. Perceptions of the severity of the crisis can actualize in two ways: perceptions of culpability and perceptions of the locus of control of the crisis. Coombs (1998) argued that if the organization is perceived by its publics as culpable, it is more likely to utilize more accommodative strategies to control the damage. Consequently, if the publics perceive that the locus of control of the crisis lies with the organization, the perception of its crisis responsibility increases. Conversely, if the publics perceive the locus of control to be external to the organization, the perception of crisis responsibility decreases.

Proposition 2: Affect—Emotional Tone, Temperature, and Importance. Jin and Cameron (2003) called for greater attention to the construct of emotion in public relations theory by providing an adapted appraisal model of emotion in public relations and a crucial dimension to the conceptualization of contingency theory. They further proposed that any given public relations stance could be assessed as a relational encounter with emotional forces that conform to a model expressed in three dimensions: first, emotional tone as the valence of the emotion ranging from negative to positive.

In measuring this level of threat consequence, it will be most proper to use a scale ranging from not negative at all to extremely negative, because threat, by nature, is negative. Emotional temperature (intensity level of the emotion) and emotional weight (importance of the emotional stimulus) for the organization and public relations professional will then be measured. Since the perceived severity works as emotional weight here, we are particularly interested in examining emotional tone and temperature as two key dimensions of the publics' emotion toward the organization in a crisis situation.

Proposition 3: Conation—Stances and Strategies. This level of consequence is directly related to what the practitioners are going to do after the above appraisals and cognitive as well as affective responses, which links well with crisis management literature regarding communication strategies (see Davis & Gilman, 2002; Lukaszewski, 1997; Ray, 1999); communications planning (Ferguson, 1999; Marra, 2004); and relationship/issues management and media relations (Cowden & Sellnow, 2002; Harrison, 1999; Massey, 2001; Ulmer, 2001; Seeger & Ulmer, 2002).

From a contingency theory perspective, Pang, Jin, and Cameron (2010) argued that the degree of accommodation or advocacy depends upon the weighing of many factors based on the matrix of factors suggested. Studies have shown that certain key variables predispose an organization to accommodation while others prohibit it. Sometimes prevailing circumstances steer the organization toward more accommodation or advocacy. Therefore, the stances adopted when facing a threat can be measured according to the movement from pure advocacy to pure accommodation.

In their analyses of how government balanced the multiple acts of managing emotions and perceptions of its multiple publics even as it attempted to deal with the evolving nature and dynamism of a health threat, namely the Severe Acute Respiratory Syndrome (SARS), Jin, Pang, and Cameron (2006) and Pang, Jin, and Cameron (2004) found that health threats seemed to be the predominant factor affecting the stances and strategies used. For instance, in their study of the Chinese government's management of SARS, the authors found that it had adopted different stances for different publics. Even as there was an overall movement from advocacy towards accommodation, the government's stances toward the different publics vary. With foreign countries/businesses, as with the general public, the government appeared to be more accommodating. With farmers and those living in rural areas, as well as those affected by SARS, including those infected, the government's stance appeared to be one of advocacy. These could be based on the nature of power relationships and the stakes involved. This is consistent with contingency theory's assertion that certain variables predisposed the organization to accommodate, namely organization's exposure, and the top management's decision power (Cameron, Pang, & Jin, 2008).

Applying the Threat Appraisal Model to Better Prepare for Crises

The Effects of Threat Appraisal in Crisis Communication. Several empirical tests have been conducted to test this threat appraisal model. For example, Pang, Jin, and Cameron (2006) adapted this model to examine the fabric and faces of threat of terrorism. The issuance of terror alerts by the Department of Homeland Security (DHS) was analyzed in terms of how the terrorism-related threat was appraised by the DHS and the conservative and liberal audiences. Findings showed a shared view by the DHS and conservative audiences while liberal audiences thought otherwise. Though there appeared to be consensus in threat communication, more internal consistency within the DHS is needed to optimize its effectiveness.

Jin and Cameron (2007) conducted an online experiment on the effects of threat type and threat duration on public relations professionals' cognitive appraisals of threats, affective responses to threats, and the stances taken in threat-embedded crisis situations. Using a 2 (external versus internal threat type) times 2 (long-term versus short-term threat duration) within-subjects design, 116 public relations professionals were exposed to four crisis situation scenarios. Research findings revealed the main effects of threat type on threat appraisal, emotional arousal and qualified-rhetoric-mixed accommodations, and the main effects of threat duration across all threat consequences. Interactions of these two threat dimensions revealed that external and long-term threat combination led to higher situational demands appraisal and more intensive emotional arousal. This study further examined the relationship between cognitive appraisal, affective responses, and stances as key aspects of threat consequences. High cognition and stronger affect regarding threats were found to be related to more accommodating stances.

Essential Role of Threat Assessment in Crisis Preparedness. Scholars agree that crisis management is a dynamic, ongoing process through a life cycle. While many scholars have posited different frameworks, in Wilcox and Cameron's (2009) proactive-strategic-reactive-recovery framework, threat assessment ought to take place in the strategic phase (i.e., before the threat erupts into a full-blown crisis). Heath (2004) emphasized the importance of strategic preparation for crisis responses because organizations are likely to suffer less damage from the crises when they are better and strategically prepared. However, there is a lack of a standard definition of crisis communication preparedness, and little research exists about the actual level of crisis preparedness (Cloudman & Hallahan, 2006). Penrose (2000) advocated the importance of examining how organizations perceive crises to understand whether their responses will be successful. Cloudman and Hallahan (2006) further identified confidence in the organization's ability to respond as a unique indicator separate from the organization's actual ability.

Taking a cognitive appraisal approach, Jin and Cameron (2007) specified resources as those that can be allocated at the moment; otherwise the crisis might be an unsolvable disaster. The threat appraisal model they posited is composed by appraisal of situational demands (including danger and uncertainty and required effort) and appraisal of allocatable organization resources (including knowledge and skill, time, finance, and the support from the dominant coalition). Previous findings suggested that organizational crisis preparedness seemed to be positively correlated to organization size, the level of autonomy, delegation of authority within the organization, and the process orientation of the organization—but not organization type (Cloudman & Hallahan, 2006). Penrose (2000) mentioned that, compared to smaller, less-known companies, huge companies with prestige might have the financial resources to better handle crisis situations.

Jin (2010) examined the perceived crisis preparedness of a random sample of public relations practitioners in the U.S. as evidenced in their assessment of crisis situation and organizational resources. The main effects of organization size and practitioner role were evident on practitioners' perceived crisis preparedness in different crisis situations. Organization type was found to be effective only when it interplayed with either practitioner role or organization size.

Using the North Korean nuclear crisis as a case study, Hwang and Cameron (2008a) found that general audience for a global crisis can become a grassroots force in the ultimate fate of policy decisions. They identified dominant coalition characteristics, external threats, and external public characteristics as overall strong predictors for public estimation about the government's stance. They further found that the general audience's perception of situational factors (external threat and external public characteristics) was a stronger predictor for the participants' stance estimation than perception of predisposing factors (dominant coalition characteristics). Considering the importance of general public support of an organization during a crisis, Hwang and Cameron (2008b) also examined how perceived leadership style influences public expectation about an organization's stance in crisis and the relationship between perceived severity of threat and the expected stance of the organization based on leadership perception. In another experimental study, Hwang and Cameron (2009) found a main effect of perceived leadership and an interaction effect of perceived leadership and perceived severity of threats on the public's estimation of organizational crisis responses. The results indicate that the contingent theoretical argument explaining the dynamics of organizational factors and situational factors that determine stance taken by public relations practices can also be applied when explaining the outside latent public's thought patterns predicting an organizational stance and message strategy. As the authors concluded, contingency theory offers predictive power—not just for the practitioner but for the public passing judgment on the stance taken by an organization.

Conclusion and Future Research Directions

A threat, without a clear idea of what it entails and how to deal with it, leads nowhere but simply arouses fear, argued Strong, Anderson, and Dubas (1993). For this chapter, we wanted to dispel any ominous notion of how threat might be perceived by attempting to explicate what threat is in a crisis, what effects it might cause, the dimensionality of threats, and a cognitive appraisal model to test the effects of threats by way of conceptualizing three testable propositions.

Jin, Pang, and Cameron (2005) conceptually differentiated threats from risk, fear, and conflict, which are the cause and the effect of crisis. They further proposed the explication of the concept by expanding, cross-fertilizing, and integrating ideas from an inter-disciplinary review of literature and enumerated the dimensionality of threats. A threat appraisal model within the contingency theory framework is based on the cognitive, affective, and conative levels of threats. Through the delineation of what threat is, we showed how threat is manifested in power relations and how this power play operates in a social structure and what options are available to address the power imbalance caused by threat. Threats in conflicts comprise two categories: threatening as well as non-threatening. While we can appreciate a greater understanding of threats in a crisis situation, to understand fully the faces and fabric of threat, we have to examine not only their antecedents but also their nature or dimensionality as well as the consequences of how an organization will react to manage the situation.

Conclusion

The threat appraisal model we propose encompasses on one level recognizing the inherent coping process that is taking place in response to threat as well as on another level—recognizing the need to appraise and distinguish the threat through the concomitant identification of primary appraisals (of situational demands) and secondary appraisals (of the individual's resources) and the interplay between demand and resource appraisals. The inter-linkages of these processes are conceptualized. Additionally, appraisal of threats necessitates an understanding of the dimensionality of threats. We operationalized dimensionality in three dimensions: threat type, threat level, and threat duration.

To fully respond to threats at these varied levels, we offer three testable propositions based on cognitive, affective and conative levels. In explicating threat, it is our intention to break down the faces and fabric of threat. At the same time, we offer a way of understanding threat and a possible model in which we respond to threat as public relations practitioners. A rooted conceptual understanding of threat in crisis will advance the theory development in public relations, in particular, better understanding and diagnosing threats in crisis, which, in turn, affects how crisis can be managed with optimal effectiveness for all concerned.

References

Benoit, W. L. (1995). A theory of image restoration. *Accounts, excuses, and apologies: A theory of image restoration discourse* (pp. 63–98). Albany: State University of New York.

———. (2004). Image restoration discourse and crisis communication. In D. Millar & R. L. Heath (Eds.), *Responding to crisis: A rhetorical approach to crisis communication* (pp. 263–280). Mahwah, NJ: Lawrence Erlbaum.

Benoit, W. L., & Pang, A. (2008). Crisis communication and image repair discourse. In T. Hansen-Horn & B. Neff (Eds.), *Public relations: From theory to practice* (pp. 244–261). Boston, MA: Pearson Allyn & Bacon.

Blascovich, J., & Mendes, W. B. (2000). Challenge and threat appraisals: The role of affective cues. In J. P. Forgas (Ed.), *Feeling and thinking* (pp. 59–82). Cambridge, UK: Cambridge University Press.

Blascovich, J., Mendes, W. B., Hunter, S. B., & Salomon, K. (1999). Social "facilitation" as challenge and threat. *Journal of Personality and Social Psychology, 77*(1), 68–77.

Cameron, G. T., Pang, A., & Jin, Y. (2008). Contingency theory: Strategic management of conflict in public relations. In T. Hansen-Horn & B. Neff (Eds.), *Public relations: From theory to practice* (pp. 134–157). Boston, MA: Pearson Allyn & Bacon.

Cancel, A. E., Mitrook, M. A., & Cameron, G. T. (1999). *Testing the contingency theory of accommodation in public relations.* Public Relations Review, 25*(2), 171–197.*

Carver, C. S. (1977). Self-awareness, perception of threat, and the expression of reactance through attitude change. *Journal of Personality, 45*, 501–512.

Cloudman, R., & Hallahan, K. (2006). Crisis communications preparedness among U.S. organizations: Activities and assessments by public relations practitioners. *Public Relations Review, 32*, 367–376.

Cohen, A. R. (1959). Situational structure, self-esteem, and threat-oriented reactions to power. In D. Cartwright (Ed.), *Studies in Social Power* (pp. 35–52). Ann Arbor, MI: Institute for Social Resarch.

Coombs, W. T. (1998). An analytic framework for crisis situations: Better responses from a better understanding of the situation. *Journal of Public Relations Research, 10*(3), 177–191.

———. (2001). Teaching the crisis management/communication course. *Public Relations Review, 27*(1), 89–109.

Corneille, O., Yzerbyt, V. Y., Rogier, A., & Buidin, G. (2001). Threat and the group attribution error: When threat elicits judgments of extremity and homogeneity. *Personality and Social Psychology Bulletin, 27*, 427–446.

Cowden, K., & Sellnow, T. L. (2002). Issues advertising as crisis communication: Northwest Airlines' use of image restoration strategies during the 1998 pilots' strike. *Journal of Business Communication, 39*(2), 193–219.

Davis, S. C., & Gilman, A. D. (2002). Communications coordination. *Risk Management, 49*(8), 38–44.

Department of Homeland Security. Retrieved November 1, 2004, from www.dhs.gov

Fearn-Banks, K. (2007). *Crisis communications: A casebook approach* (3rd ed.). Mahwah, NJ: Lawrence Erlbaum.

Ferguson, S. D. (1999). *Communication planning*. Thousand Oaks, CA: Sage.

Grunig, J. E., & Grunig, L. A. (1992). Models of public relations and communications. In J. E. Grunig (Ed.), *Excellence in public relations and communication management* (pp. 285–326). Mahwah, NJ: Lawrence Erlbaum.

Grunig, J. E., & Huang, Y. H. (2000). From organizational relationship to relationship indicators: Antecedents of relationships, public relations strategies and relationship outcomes. In J. Ledingham & S. D. Bruning (Eds.), *Public relations as relationship management* (pp. 23–53). Mahwah, NJ: Lawrence Erlbaum.

Harrison, S. (Ed.). (1999). *Disasters and the media*. London: Macmillan.

Heath, R. L. (2004). Crisis preparation: Planning for the inevitable. In D. P. Millar and R. L. Heath (Eds.), *Responding to crisis: A rhetorical approach to crisis communication* (pp. 33–35). Mahwah, NJ: Lawrence Erlbaum.

Hwang, S. W., & Cameron, G. T. (2008a). The elephant in the room is awake and takes things personally: The North Korean nuclear threat and the general public's estimation of American diplomacy. *Public Relations Review, 34*, 41–48.

———. (2008b). Public's expectation about an organization's stance in crisis communication based on perceived leadership and perceived severity of threats. *Public Relations Review, 34*, 70–73.

———. (2009). The estimation of a corporate crisis communication. *Public Relations Review, 35*, 136–138.

Jin, Y. (2010). The interplay of organization type, organization size and practitioner role on perceived crisis preparedness: A cognitive appraisal approach. *Journal of Contingencies and Crisis Communication, 18*(1), 49–54.

Jin, Y., & Cameron, G. T. (2003). *Rediscovering emotion in public relations: An adapted appraisal model and an emotion-laden contingency plane.* Unpublished manuscript.

———. (2007). The effects of threat type and duration on public relations professionals' cognitive, affective and conative responses in crisis situations. *Journal of Public Relations Research, 19*(3), 255–281.

Jin, Y., Pang, A., & Cameron, G. T. (2006). Do we stand on common ground? A threat appraisal model for terror alerts issued by the Department of Homeland Security. *Journal of Contingency and Crisis Management, 14(2),* 82–96.

———. (2005). *Explicating threats: Towards a conceptual understanding of the faces and fabric of threat in an organizational crisis.* New York City, NY: ICA.

———. (2004). *"SARS-free, now to fine-tune crisis handling": A contingency approach to analyzing the Singapore government's management of the perception and emotion of its multiple publics*. New Orleans, LA: ICA.

Lanzetta, J. T., Haefner, D., Langham, P., & Axelrod, H. (1954). Some effects of situational threat on group behavior. *Journal of Abnormal and Social Psychology, 49,* 445–453.

Lukaszewski, J. E. (1997). Establishing individual and corporate crisis communication standards: The principles and tools. *Public Relations Quarterly, 42*(3), 7–15.

Marra, F. J. (2004). Excellent crisis communication: Beyond crisis plans. In D. P. Millar & R. L. Heath (Eds.), *Responding to crisis: A rhetorical approach to crisis communication* (pp. 311–326). Mahwah, NJ: Lawrence Erlbaum.

Maslow, A. H. (1943). Conflict, frustration, and the theory of threat. *Journal of Abnormal and Social Psychology, 38*, 81–86.

Massey, J. E. (2001). Managing organizational legitimacy: Communication strategies for organizations in crisis. *The Journal of Business Communication*, *38*(2), 153–182.

Mulder, M. & Stemerding, A. (1963). Threat, attraction to group, and need for strong leadership. *Human Relations, 16*(4), 317–334.

Oppe, S. (1988). The concept of risk: A decision theoretic approach. *Ergonomics, 31*(4), 435–440.

Pang, A. (2008). Contingency model of conflict. In W. Donsbach (Ed.), *International Encyclopedia of Communication* (pp. 978–980). Malden, MA: Blackwell.

Pang, A., Jin, Y., & Cameron, G. T. (2004). *"If we can learn some lessons in the process": A contingency approach to analyzing the Chinese government's management of the perception and emotion of its multiple publics during the Severe Acute Respiratory Syndrome (SARS) crisis*. Miami, FL: IPPRC.

———. (2006). Do we stand on common ground? A threat appraisal model for terror alerts issued by the Department of Homeland Security. *Journal of Contingency and Crisis Management, 14*(2), 82–96.

———. (2010). Contingency theory of strategic conflict management: Directions for the practice of crisis communication from a decade of theory development, discovery and dialogue. In W. T. Coombs & S. J. Holladay (Eds.), *Handbook of crisis communication* (pp. 527–549). Malden, MA: Wiley-Blackwell.

Penrose, J. M. (2000). The role of perception in crisis planning. *Public Relations Review, 26*(2), pp. 155–171.

Ray, S. J. (1999). *Strategic communication in crisis management*. Westport, Connecticut: Quorum.

Richards, B. (2004). Terrorism and public relations. *Public Relations Review, 30*, 169–175.

Rosenthal, U., & Kouzmin, A. (1997). Crises and crisis management: Toward comprehensive government decision-making. *Journal of Public Administration Research and Theory, 7*(2), 277–305.

Seeger, M. W., & Ulmer, R. R. (2002). A post-crisis discourse of renewal: The cases of Malden Mills and Cole Hardwoods. *Journal of Applied Communication Research, 30*(2), 126–138.

Strong, J. T., Anderson, R. E., & Dubas, K. M. (1993). Marketing threat appeals: A conceptual framework and implications for practitioners. *Journal of Managerial Issues, 5*(4), 532–546.

Ulmer, R. R. (2001). Effective crisis management through established stakeholder relationships. *Management Communication Quarterly, 14*(4), 590–615.

Volkan, V. (1997). *Bloodlines: Ethnic pride and ethnic terrorism*. New York: FSG.

Wilcox, D. L., & Cameron, G. T. (2009). *Public relations strategies and tactics* (9th ed.). NY: Pearson Allyn & Bacon.

Wrigley, B., Salmon, C. T., & Park, H. S. (2003). Crisis management planning and the threat of bioterrorism. *Public Relations Review, 29*, 281–290.

Appendix 1: Table of Variables That Affect an Organization's Response

Internal Variables

Organization Characteristics

- Open or closed culture
- Dispersed widely geographically or centralized
- Level of technology the organization uses to produce its product or service
- Homogeneity or heterogeneity of officials involved
- Age of the organization/value placed on tradition
- Speed of growth in the knowledge level the organization uses
- Economic stability of the organization
- Existence or non-existence of issues management officials or program
- Organization's past experiences with the public
- Distribution of decision making power
- Formalization: number of roles or codes defining and limiting the job
- Stratification/Hierarchy of positions
- Existence or influence of legal department
- Business exposure
- Corporate culture

Public Relations Department Characteristics

- Number of practitioners total and number of college degrees
- Type of past training: Trained in PR or ex-journalists, marketing, etc.
- Location of PR department in hierarchy: Independent or under Marketing umbrella/experiencing encroachment of marketing/persuasive mentality
- Representation in the Dominant Coalition
- Experience level of PR practitioners in dealing with crisis
- General communication competency of department
- Autonomy of department
- Physical placement of department in building (near CEO and other decision makers or not)
- Staff trained in research methods
- Amount of funding available for dealing with external publics
- Amount of time allowed to use dealing with external publics
- Gender: percentage of female upper-level staff/managers
- Potential of department to practice various models of public relations

Characteristics of Dominant Coalition (Top Management)

- Political values: Conservative or liberal/open or closed to change
- Management style: Domineering or laid-back
- General altruism level
- Support and understanding of PR
- Frequency of external contact with publics
- Departmental perception of the organization's external environment
- Calculation of potential rewards or losses using different strategies with external publics
- Degree of line manager involvement in external affairs

Internal Threats (How Much is at Stake in the Situation)

- Economic loss or gain from implementing various stances
- Marring of employees' or stockholder's perception of the company
- Marring of the personal reputations of the company decision makers

Individual Characteristics (Public Relations Practitioners, Domestic Coalition, and Line Managers)

- Training in diplomacy, marketing, journalism, engineering, etc.
- Personal ethics
- Tolerance or ability to deal with uncertainty
- Comfort level with conflict or dissonance
- Comfort level with change
- Ability to recognize potential and existing problems
- Extent to openness to innovation
- Extent to which individual can grasp other's worldview
- Personality: Dogmatic, authoritarian
- Communication competency
- Cognitive complexity: Ability to handle complex problems
- Predisposition toward negotiations
- Predisposition toward altruism
- How individuals receive, process and use information and influence
- Familiarity with external public or its representative
- Like external public or its representative
- Gender: Female versus male

Relationship Characteristics

- Level of trust between organization and external public
- Dependency of parties involved
- Ideological barriers between organization and public

External Variables

Threats

- Litigation
- Government regulation
- Potentially damaging publicity
- Scarring of company's reputation in business community and in the general public
- Legitimizing activists' claims

Industry Environment

- Changing (dynamic) or static
- Number of competitors/level of competition
- Richness or leanness of resources in the environment

General Political/Social Environment/External Culture

- Degree of political support of business
- Degree of social support of business

The External Public (Group, Individual, etc.)

- Size and/or number of members
- Degree of source credibility/powerful members or connections
- Past successes or failures of groups to evoke change
- Amount of advocacy practiced by the organization
- Level of commitment/involvement of members
- Whether the group has public relations counselors or not
- Public's perception of group: reasonable or radical
- Level of media coverage the public has received in past
- Whether representatives of the public know or like representatives of the organization
- Whether representatives of the organization know or like representatives from the public
- Public's willingness to dilute its cause/request/claim
- Moves and countermoves
- Relative power of organization
- Relative power of public

Issue Under Question

- Size
- Stake
- Complexity

Chapter 8

Emergency Management Planning: Risk, Crises, Issues, and Social Media

Michael J. Palenchar & Karen Freberg

The authors tackle a very timely and pertinent issue by exploring the role social media have in emergency management planning. Popular media recognizes how social media are employed in the midst of a crisis situation. This chapter significantly expands our understanding of that relationship as the authors detail how it can be utilized in preparation measures. They wisely acknowledge the dangers here and advocate a strong risk assessment before establishing social media relationships. That is followed with insights into how a social media presence can be established, maintained, and used to empower stakeholders.

Introduction

Increasingly important and explosively growing new media technology within crisis communication are early warning and emergency response measures through social media, within the context and application of issues management and risk communication, as fundamental elements of pre-crisis communication and management. Two of the key questions within crisis communication are *what needs to be communicated* and *how should it be communicated prior to a crisis?* Another key question is *what is the relationship of a risk evolving into a crisis and quite possibly a long-term issue that needs to be managed to better understand the original and new risks?* Fundamentally, this requires an early forecasting system and a proactive public relations and risk communication program. Successful communication might or might not prevent a crisis, but it absolutely can lay a foundation for managing crisis responses should a risk become manifested into a crisis and potentially a long-term issue to be managed by the risk generator and risk-bearing stakeholders within a community. The purpose of this chapter is to demonstrate social media as an emerged and important element within emergency management and crisis communication planning and to address several notable aspects of using social media in pre-crisis communication, including issues related to the explosion of mobile technology, training, speed and consistency, control, scanning and monitoring, risk assessment, transparency, prior engagement, and vulnerable populations.

Social Media at the Intersection of Risk, Crises, and Issues Communication

Risk Communication

Recent approaches to risk communication have reinforced the importance of a dialogic, relationship-building approach to managing risk concerns and meanings of community residents. In this approach, information does not completely account for the final estimation of risk. Culture, values, attributions of responsibility, and self-interests employed in risk interpretations may be more important than technical data (Dake, 1992; Renn, Burns, J. Kasperson, R. Kasperson, & Slovic, 1992; Vaughan & Seifert, 1992).

Strategic risk communication charges that each key public makes an idiosyncratic reaction to each risk based on its distinct yet continually transforming decision heuristic. Risk-relevant publics exist because people have different interpretive heuristics as well as conflicting understanding of whether something creates risk, whether that risk should be tolerated, and whether avoidance strategies or control measures are warranted.

This perspective has been characterized as an infrastructural approach to risk communication (e.g., Heath, 1996; Palenchar & Heath, 2002), one that acknowledges that political dimensions of risk communication couple with the more purely technical or scientific dimensions. In a community where varied opinions and concerns exist, scientific assessment of facts and performance achievements are subjected to persuasive contests and sociopolitical interpretations that include consideration of balances of political power (Heath & Nathan, 1991).

The assumption of an infrastructural approach to risk communication is that even though sound science is critical, the myriad of voices will offer various interpretations of that science. According to Heath and Palenchar (2009) the infrastructural theory works to embrace the need for voices to produce facts, engage in evaluations, foster identifications, and develop policies that work best within each society to define, control, accommodate, and otherwise manage risks or risk responses. "The challenge is to participate in and even create infrastructures that allow multiple voices to speak on risks and develop culturally sensitive approaches that evolve through dialogue rather than are prescribed by the assumed authority of one or more individual organizations or institutions" (p. 326). An infrastructural approach is central to the understanding of risk democracy, which reasons that in association with various experts, individual voices of risk bearers must be brought together to create systems and shared perspectives that appropriately assess, mitigate, and respond to risks that may be unevenly borne throughout any community or larger society.

In their struggle to control risks, according to Heath and Palenchar (2009), people seek and contest facts, evaluative premises, and draw policy conclusions to be derived from those facts and premises. They also create identifications, whether they are professional, organizational, or of other kinds, including issue-oriented activists. "Residents of every community, culture, and country have legitimate reasons to be concerned for their own and others' health, safety, and environmental quality, especially since there are no absolute standards or, as some would argue, no objective standards" (p. 312). Heath and Palenchar suggested that risk communication and issues management merge at the point where key publics feel deeply that organizations create or allow risks to occur that will affect the health, safety, environmental quality, and economic well-being of community residents and users of products.

Decision-making Aimed at Control. Within risk communication, key factors in assessing and communicating about risk are the differences between one's own perceptions of ability to control a risk relevant to one's own interest, as opposed to the perception of how capable others are to control risks relevant to their behavior. One of the keys in community risk perception studies is the tendency of potential risk bearers, risk generators, and policy makers to be inaccurate in their ability to assess properly the likelihood that certain risk events will occur and, if they occur, whether they will lead to negative or positive outcomes (Heath & Palenchar, 2009). The issue to be managed is how to understand and control risks as well as gain acceptance for these measures in ways that foster the wisest outcomes in any community of interest (Freudenberg, 1984).

An issues management approach to risk assessment and communication argues that these activities are more likely to be successful when they increase the control that key publics have over the technical information, risk assessment process, and regulation of any threatening product as well as its manufacture or transportation. Control is perceived to work best when it arises from a constructive partnership between organizations that create risk and local government that champions the interests of citizens (Heath, Bradshaw, & Lee, 2002).

Issues Management

According to Heath and Palenchar (2009), from a broad perspective, issues management is the amalgamation of organizational functions and responsive culture that brings together strategic business planning, issue monitoring, best-practice standards of corporate responsibility, through dialogue to foster a supportive climate between each organization and stakeholders. From a societal view that features the logics of power resource management, and according to Heath (2006), issues management can be defined as "the management of organizational and community resources through the public policy process to advance organizational interests and rights by striking a mutual balance with those of stakeholders" (p. 79).

Issues management "is a leadership process that defines the strategic common ground between a company [or other organization] and its key audiences" (Palese & Crane, 2002, p. 284). The role of the communication professional may best be an advocate for discussions, decisions, communications, and behaviors that fall in the area of doing the right thing (Berger & Reber, 2006). The challenge is to engage in the contest of what constitutes the right or best thing while managing risk and crisis information.

For many scholars, legitimacy is at the heart of issues management (Heath & Palenchar, 2009; Roper, 2002, 2005; Spencer, 2004). Fleisher (2001) observed that publics in general and around the world tend to doubt that corporate interests correspond to theirs. Issues management is a clash for legitimacy and thus power. Efforts to define issues management must not lose sight of organizations' need to harmonize organizational and stakeholder interests, especially during pre-crisis communication.

To achieve its potential, according to Heath and Palenchar (2009), issues management must add value by allocating, defining, and distributing resources: human, financial, and material within emergency management planning, including pre-crisis communication efforts. Although competing and conflicting interests are such that all cannot be equally satisfied, issues management serves best when it assists in the planning, analysis, communication, and coalition-building efforts by which mutual interests are sought and appropriate resource allocation is achieved for managing crisis events. Part of these efforts is the consideration of stakeholders' perceptions and meaning of risk associated with pre-crisis planning.

According to Mitchell, Agle, and Wood (1997), "Persons, groups, neighborhoods, organizations, institutions, societies, and even the natural environment are generally thought to qualify as actual or potential stakeholders" (p. 855). This line of research, which partially can be attributed to the 1984 publication of Freeman's *Strategic Management: A Stakeholder Approach*, adds value when it helps the organization identify stakeholders and segment different kinds of publics from stakeholder categories (Grunig, 2006) to improve communication.

Not surprisingly then, stakeholders are socially constructed entities that are integral components of the issues management and risk and crisis communication process. Since organizations must address multiple stakeholders and their concerns at any given time, the effort to identify and prioritize those stakeholders on which the organization should focus its attention is known as stakeholder theory (Bridges, 2004; Mitchell et al., 1997).

Pre-crisis Communication

While significant amounts of literature have focused on the messages to communicate during a crisis and how to recover from the crisis, more research is needed to determine what needs to be done before a crisis occurs. Jaques (2010) stated that crisis communication should put more emphasis on the "management activities proceeding from potential crisis identification and prevention through event response and on to long term post-crisis management" (p. 469).

One of the critical features of planning for a crisis is the development of pre-crisis communication tactics. Samples of traditional pre-crisis communications consist of templates for messages from top spokespersons, dark websites, and draft press releases. In an environment of social media and mobile devices, pre-crisis planning must expand beyond to include messages appropriate to these platforms. Individuals affected by a crisis are more likely today to search for information using the Internet and new, mobile media applications (Stephens & Malone, 2009).

One of the challenges for the field of crisis communications is the difficulty of identifying and preparing for the "low-probability, high-impact event" that can lead to a crisis (Pearson & Clair, 1998, p. 60). Crisis situations are exceptional events that do not happen on a regular basis (Roux-Dufort, 2007). Examples of recent exceptional events precipitating crisis situations include natural disasters like the H1N1 global public health crisis in 2009 and the devastating earthquake in Haiti in 2010. Crisis communication professionals strategize and prepare for all (most!) unexpected events that have the potential to occur. These events can take the form of a one-time incident, such as a terrorist attack, or an ongoing process, such as a flu pandemic (Jaques, 2010).

To identify potential crises, possible major events should be considered, but smaller issues and risks that could transform into larger issues should not be overlooked. Anticipating types of crises ahead of time and categorizing them based on their salient characteristics is essential for effective preparation. Categorization of potential crises can provide guidance for their management. Effective crisis communication practices take into account that different types of crises could potentially occur at any given point in time and recognize that each type is associated with its own challenges and risks for crisis communication professionals and the organizations they represent in a time of emergency. At the same time resources need to be spent on preparing for crises that defy traditional experience and research and are also unpredictable.

Once potential crises have been identified and categorized, pre-crisis planning can provide a foundation for rapid and effective responses. Some case studies point to more effective preparation. Most public relations practitioners have indicated that their organizations prepare crisis communication plans (Cloudman & Hallahan, 2006). Although Hurricane Katrina has been viewed as one the most devastating natural disasters in United States' history, many effective and proactive crisis communication strategies were already in place. The Centers for Disease Control and Prevention (CDC) had prepared for hurricanes by reviewing their crisis materials and ensuring that their website was up to date with the latest information for users to access, such as hurricane readiness, drowning prevention, shelter information, and food safety information. In addition, the CDC prepared their hurricane-related crisis communication messages not only in English but also in Spanish, Vietnamese, French, Portuguese, and German (Vanderford, Nastoff, Telfer, & Bonzo, 2007). Also, the public health industry demonstrated effective preparation during the avian flu pandemic in 2004 (Avery & Kim, 2009); press releases from the top four heath organizations (Centers for Disease Control and Prevention (CDC), United States Department of Health and Human Services (HHS), National Institutes for Health (NIH), and World Health Organization (WHO)) about the avian flu demonstrated that public information officers understood the importance of strategic public partnerships in preparing for a pandemic health crisis.

An example of the importance of including social media in crisis communications occurred during the 2010 Haiti earthquake. This crisis highlighted a more mature use of signal messages services (SMS) text messages to communicate first response aid to individuals needing immediate medical attention or who were trapped under buildings and other fallen structures. Mobile phones were used to communicate first aid information and to provide information about where to go for shelter, food, water, and other health assistance (Bulkely, 2010). Examples of some of the messages that were being sent via these mobile devices included information for medical care ("Hospital Sacre-Coeur in Milot says it has capacity for patients and asks people to make their own way there"); search and rescue ("Though the government says the search and rescue phase is over, SAR teams are still available. If you know someone is trapped call + 870 764 130 944, email haiti.opc@gmail.com or contact MINUSTAH"); and general advisories on other issues of relevance ("Information in a crisis," June 18, 2010, para.1–2). The growing prevalence of mobile phone ownership and use makes rescue efforts possible that would be unthinkable ten years ago.

What is Social Media?

The role and use of social media during a crisis just beginning to be understood. Numerous definitions of social media exist, and consensus across disciplines has not yet been reached. Social media combine an eclectic range of online; word-of-mouth forums including blogs, discussion boards and chat rooms; consumer-to-consumer e-mail; consumer product or service ratings websites and forums; Internet discussion boards and forums; and microblogs (Mangold & Faulds, 2009). Lariscy, Avery, Sweetser, and Howes (2009) defined social media as "online practices that utilize technology and enable people to share content, opinions, experiences, insights, and media themselves" (p. 1). Social media provides several functions, including creation and participation, openness, dialogue, and community.

Some researchers and practitioners argue that social media is revolutionary in its effects on how organizations and individuals communicate with each other (Solis, 2008a). Social media in general is "much more than user-generated content. It's driven by people in the communities where they communicate and congregate. They create, share, and discover new content without our help. They're creating vibrant and rich cultures across online networks and using the social tools that we learn about each and every day to stay connected" (p. 5).

Another term related to social media is social networking sites (SNS). Whereas social media focus on establishing a specific platform for individuals to communicate with each other, SNS can be defined as web-based services that allow individuals to (1) construct a public or semi-public profile within a bounded system, (2) articulate a list of other users with whom they share a connection, and (3) view and traverse their list of connections and those made by others within the system (Boyd & Ellison, 2008). One of the primary functions of SNS is to establish relationships (Waters, Burnett, Lamm, & Lucas, 2009), whereas social media is more about "people" in general (Marken, 2007). SNSs typically differ from previous online technologies because individuals connect with those with whom they have an existing offline relationship rather than seeking to create new relationships with people whom they do not know in person (Ellison, Steinfield, & Lampe, 2007).

Some types of social media are more popular with corporations and other organizations, whereas some are more frequently used for personal use. Four types of social media are most frequently used among business professionals and corporations: LinkedIn, Facebook, Twitter, and blogs (Gordon, 2009). Blogs in particular have been used to communicate and "initiate and manage public engagement in corporate communications" (Yang & Kang, 2009, p. 323). Social media offers many benefits for organizations. It establishes credibility for the brand and overall reputation among its key publics (Prentice & Huffman, 2008). Marken (2007) stated that social media can provide "an unfiltered view of consumer perceptions so firms can see what will impact the future of their business; word-of-mouth is having a tremendous control over perception and acceptance, and will become increasingly significant in influencing companies and products according to the revolutionaries" (p. 10).

Social Media and Crisis Communications

> The social web is creating a fundamental shift in disaster response—one that will ask emergency managers, government agencies, and aid organizations to mix time-

> honored expertise with real-time input from the public....We need to work together to better respond to that shift. (American Red Cross, 2010, para.6)

The ever-changing global environment with its possibility of life-altering crises has elevated the awareness of the importance of pre-crisis communication among business and public relations professionals. Businesses invest time, money, and public relations resources into the latest forms of technology, including the use of social media, to ensure that they are proactive and prepared to address potential crises that may have consequences to their financial health or their reputation among target audiences.

Given the effectiveness of mobile and social media, organizations are well advised to include the preparation of pre-crisis communications tailored to these platforms that are analogous to more traditional pre-crisis communications of statements, press releases, and dark websites. What is needed is more research to identify the most effective practices, enabling successful planning.

Gaps exist in our understanding of effective pre-crisis communication practices, especially given the advent of social media. Stephens and Malone (2009) reported that people in a crisis situation wanted the opportunity to engage in dialogue and share information (website links) with others. By being aware of these trends related to new media, crisis communication professionals can adapt the format of their crisis communication messages to new media to fit these expectations of their key stakeholders (Stephens & Malone, 2009). Social media do not require crisis communicators to reinvent the wheel, but, rather, to adapt what they already know to the new platforms. "The most robust solution is to combine the best of what we know about crisis communication in emergency situations with an increasing understanding of how distributed networks, based on social media platforms, can become a viable and effective toolkit" (Semple, 2009, para. 10).

Previous research can guide investigations into the social and mobile media parallels of established practices. For example, crisis communication messages are most effective if they have a personal touch as well as provide individuals with action steps to take (Dillard, Shen, & Vail, 2007; Heath, Lee, & Ni, 2009). Research could be conducted to identify how the personal touch is affected by media platforms, including the use of text, voice, or video.

The crisis communication literature also emphasizes the need for established partnerships that are set in place before a crisis occurs. With the immediacy provided by social and mobile media, coordination of messages among relevant organizations becomes even more critical than before. In an analysis of the response to Hurricane Katrina, Vanderford et al. (2007) noted a need for more integration and partnerships (i.e., weathercasters and meteorologists) in working together to disseminate information immediately to audiences impacted by the hurricane. Avery and Kim (2009) found in their study looking at press releases on the avian flu that the public information officers involved in the press releases were proactive in communicating the importance of having these partnerships and coordinating their efforts with other organizations, particularly with other government organizations. Coordinated dissemination of information via social and mobile media is likely to require very strong partnerships, and further research can illuminate how this might be best accomplished.

One recent example demonstrates the possible role of social media during a risk or crisis event. On January 1, 2009, a US Airways flight was heading from New York to North Carolina when it was forced to make a crash landing into the Hudson River. Janis Krums, a citizen on a ferry, took a picture of the plane with his iPhone and uploaded it to TwitPic (a mobile photosharing site that posts directly to Twitter). Within three hours the photo was viewed online 40,000 times and was seen on several national news networks and newspapers (Terdiman, 2009). This is the impact of social media—intensified by mobile technology—to collect and spread instantaneous information. In the absence of pre-crisis communication platforms for dialogue, those who experience risk will oftentimes find an online social media forum for their discussions.

A few researchers have addressed the issue of how social media has impacted the fields of both crisis communications and public relations. First, most of the research studies have focused on studying privacy and impression management (Hinduja & Patchin, 2008; Krämer & Winter, 2008; Walther, 2007; Walther, Van Der Heide, Kim, Westerman, & Tong, 2008) or have looked specifically at social networking profiles (e.g., Boyd & Ellison, 2008; Krämer & Winter, 2008; Zhao, Grasmuck, & Martin, 2008). Others have focused primarily on social networking sites like Facebook or MySpace (Boyd & Ellison, 2008; Palen, Vieweg, Sutton, Liu, & Hughes, 2007; Steinfield, Ellison, & Lampe, 2008; Vieweg, Palen, Liu, Hughes, & Sutton, 2008) and how individuals are constructing online identities with social media (Mazer, Murphy, & Simonds, 2007; Zhao et al., 2008).

Numerous research studies and white papers have been written on social media by public relations practitioners and scholars (Kelleher, 2008; Wright & Hinson, 2008, 2009). Topics include investigations into how public relations uses social media in communicating with journalists (Laricy et al., 2009), the potential risks that public relations professionals have to address online (Pavlik, 2008), and how crisis communicators are handling a crisis with social media (Coombs, 2007, 2008; Palen et al., 2007; Vieweg et al., 2008).

Clearly, social media is a developing focus for risk and crisis communication. Numerous meetings, such as the International Conference on Information Systems for Crisis Response and Management, held in Washington, D.C., in May 2008; the Expert Roundtable on Social Media and Risk Communication During Times of Crisis at the American Public Health Association headquarters in Washington, D.C., in March 2009; and the recent Emergency Social Data Summit in August 2010 in Washington, D.C. demonstrates this as well.

Government, industry, trade, and university research reports demonstrate the explosive growth of Web 2.0 or social media in all levels of the private and public sector. For example, according to a report issued by the Chief Information Officers Council (CIO Council) of the U.S. federal government, "The use of social media for federal services and interactions is growing tremendously, supported by initiatives from the administration, directives from government leaders, and demands from the public" (p. 6). On January 21, 2009, President Barack Obama signed the Transparency and Open Government memorandum, which utilizes Web 2.0 technologies to engage with the public.

Overall, social media characteristics are about openness, conversation and dialogue, relationship development, multiple voices, and getting the message to stakeholders. A 2010 Pew Internet study showed that 82% of internet users (representing 61% of all American adults) looked for information or completed a transaction on a government website in the past year, with 25% getting advice or information from a government agency about a health or safety issue (Smith, 2010). The study also showed how citizens are organized around new online platforms and beyond websites, with nearly one-third (31%) of online adults using platforms such as blogs, social networking sites, online video, text messaging, and portable digital devices. With this basic understanding of social media and risk, crisis, and issues communication and management, several specific challenges pertain to pre-crisis communication.

Explosion of Mobile Technology. Among the relevant communication technologies are digital hand-held devices, including mobile phones, personal digital assistants, and wireless tablets such as the iPad. The evolution of the mobile phone device has taken center stage in the communication technology realm, shaping the network communications framework and the ways we connect with each other. Professional predictions point to the remarkable power of the hand-held device in the future (Baekel, 2008). According to Baekel, the very near future will see social networking done entirely on these mobile devices as opposed to static workstations, and up to 80 percent of Internet traffic will occur on mobile phones or other transferable devices. According to O'Dell (2010), mobile web will be the dominant force for obtaining information no later than 2015. A Morgan Stanley analyst suggested that the world is currently in the midst of the fifth major technology cycle of the past half-century, predicting that within the next five years more users will connect to the Internet over their mobile devices than desktop PCs (Ingram, 2010).

However, new technologies allow the entire online community—both domestically and internationally—to obtain information that can potentially create more problems for those tasked with managing a crisis. Officials might experience "sousveillance," in which bystanders use their phones to record video or take photos of emergency personnel who are not acting professionally (Stephenson & Bonabeau, 2007). Under the stress of a crisis, the immediacy of digital communication might result in false information being communicated to stakeholders (Vieweg et al., 2008). In addition, stakeholders can use digital technologies to create and disseminate their own influence, de-centralize the dissemination of information, and reduce official control.

There are other concerns with the use of hand-held digital devices as a fundamental communication infrastructure in risk and crisis situations. Traynor's (2008) research about the technical capabilities of emergency text messages as part of emergency alert systems (EAS) brings up issues with adopting technology without thinking about such risks. As university and other organizations have subscribed to more and more of these text messaging EAS systems, there are problems with technology, speed of information, systems overloads, and delivery of critical information between emergency responders or the public to 911 services.

Mobile devices implemented in emergency communications plans will allow a more even distribution and coordinated effort that will be beneficial for all parties involved in the disaster situation. Emergency managers can be actively communicating via text message or post other information (e.g., videos and photos) to establish credibility and their authority as a primary source of information in the disaster situation for impacted communities. Connecting online and using these mobile devices effectively will establish a stronger virtual community that will be more informed and engaged in the disaster recovery and implementation process (Sutton, 2010). Unfortunately, emergency management professionals appear to be lagging behind in the implementation of mobile media in their communication plans (Nichols, 2010).

Training. Organizations need to provide training as well. Cloudman and Hallahan (2006) analyzed public relations practitioners training in crisis communication and found a lack of emphasis on overall training. Most crisis and risk communication training focused on briefings and spokesperson training but did not take advantage of training with other techniques or technologies. For example, professionals using mobile devices in a crisis or disaster situation must be prepared with adequate training. Otherwise, they might inadvertently consume the entire available bandwidth or cause a complete network crash (Stephenson & Bonabeau, 2007).

The Chief Information Officers (CIO) Council (2009) argued for an increase in training since "users are almost always the weakest link in an information system, and may inadvertently divulge sensitive information through a social network" (p. 14). Managing multiple communication channels, deciding what information should be disseminated to which audiences, and preparing the best strategies for rapidly changing emergency situations require careful and extensive planning.

Speed, Consistency, and Information Overload. Reynolds and Seeger (2005) stated that crisis communication "seeks to explain the specific event, identify likely consequences and outcomes, and provide specific harm-reducing information to affected communities in an honest, candid, prompt, accurate, and complete manner" (p. 46). Effective crisis communication is quick, accurate, and consistent (Coombs, 2007). In addition, Coombs advises that crisis communication should emphasize the importance of public safety and use multiple communication channels (the Internet, Intranets, and other mass communication systems). Organizations have to be transparent and operate with a 24/7 mentality and recognize their role in the international digital business community (Heitmann & Lott, 2008; Waters et al., 2009). To survive a crisis in the age of social media, professionals working in pre-crisis communication have to be aware of the potential of their stakeholders receiving false information regarding the situation and potential rumors (Vieweg, et al., 2008).

Among the challenges facing victims of a disaster is the risk of information overload due to extensive coverage available in both traditional and mobile media (Cordner & Scarborough, 2010). Some researchers argue that crisis managers could benefit from taking the need-to-know approach, in which information designed to reduce victims' levels of uncertainty becomes a priority over general information that can be shared with others online or in person (Cordner & Scarborough, 2010). Several other important issues arise as well. One of the challenges facing public relations is that there are so many new channels and platforms of social media emerging rapidly that it is hard to get a handle on which ones are relevant to the profession (McKay, 2009).

However, social media remains a communication vehicle, one with an ambitious agenda, but the power remains in the communicator or communicating organization and their behaviors and narrative content—not in the technology. "The real value of any communication—social media included—remains the quality of the content being disseminated around the actions a brand or company is taking, the empathy for affected stakeholders being displayed, and the appropriateness and relevance of the context and perspective being provided" (Aherton, 2009, p. 3). What is scary for many organizations and risk and crisis communicators is that social media speeds up communication, speeds up awareness, and often speeds up awareness of mistakes. The real question is whether it speeds up response and appropriate behavior.

Control. One significant challenge that social media throws at risk and crisis communicators is that they perceive that they have less control over the communication messages being shared. The public seems to be more fragmented in terms of which medium they are using to obtain information, which leads to a substantial barrier in a crisis (González-Herrero & Smith, 2008), as opposed to relying on large, traditional media outlets as the medium for information exchange. The traditional strategies presented in university public relations classes do not necessarily work to the corporations' advantages in social media (Deragon, 2008). For example, public relations courses have discussed a structured step-by-step approach in handling a crisis situation that is raised by a known stakeholder (e.g., activist group or employees). However, in social media where anyone can post anything about a person or corporation anonymously, this changes the playing field. Instead of having the knowledge of knowing who is causing this crisis, public relations professionals today are challenged in addressing crises that are raised by faceless entities online. Public relations firms and professionals embrace social media strategically while acknowledging that many of the traditional frameworks public relations apply will more likely not be successful when incorporating social media (Rand & Rodriguez, 2007).

Solis (2008b), author of the *PR 2.0 Web Site*, states that social media "forces companies to look outward to proactively find the conversations that are important to business and relationships. And it's not just the responsibility of PR: it requires the participation by multiple disciplines across the organization in order to genuinely provide meaningful support and information" (pp. 6–7). Recognizing the power of social media allows public relations professionals the opportunity to use social media strategically to look for potential issues relevant to their stakeholders, prepare for a crisis, develop a plan tailored to implementation online, and to make sure that they have a plan in place to respond to a crisis that came about through social media (González-Herrero & Smith, 2008, p. 144).

Environmental Scanning and Monitoring. Although the use of social media to respond to a crisis is an emerging and significant phenomenon in public relations, social media impacts other aspects of a crisis that also warrant further study. Good pre-crisis communication planning includes monitoring and planning for eventual crises. Social media can provide powerful tools for identifying crises at a very early stage, where intervention is likely to be more effective. For example, a corporation could establish a set of Google alerts, search specifically for their corporation or high-profile individual associated with the corporation on social media search engines like Samepoint, or conduct focused searches of Twitter on a regular schedule to monitor talk about the corporation and its products. In a recent example of how emerging crises might be detected early using social media, Google maintains a large monitor in its corporate lobby, refreshing regularly to show the most popular searches. One of the first hints that Mexico was entering its swine flu crisis, long before any news stories broke in the traditional media, was a flurry of searches from that country on swine flu. Establishing best practices for these monitoring activities should be a priority for public relations researchers.

Using social media, like wikis, on day-to-day projects can streamline communication and increase efficiency. For example, the document management of a crisis plan can be carried out through social media, which increases the buy-in potential for the creation and refinement of a crisis plan as well as increases the potential for interactivity and appropriate behavior in a crisis response. In addition, tracking issues through social media and providing the crisis management team with the reports can increase the potential that a crisis will be addressed sooner and demonstrate to the team why social media needs to be embraced in the crisis response. At the very least environmental scanning of Web 2.0 media is critically important as indicated by the U.S. intelligence community, who are very concerned that terrorists might use microblogging sites such as Twitter and other social media to coordinate attacks (Musil, 2008).

Risk Assessment. Crisis management also involves risk assessment, and here again social media is changing some of the rules. Risks can reside in many locations, but it can exist within social media itself. The anonymous nature of social media, where a person can start a rumor or criticize a product without any personal responsibility, is unlike the world of the traditional media in which a person usually signs a name to a letter to the editor. When people are anonymous, their behavior can become quite negative, and this is a process known in social psychology as de-individuation (Festinger, Pepitone, & Newcomb, 1952). Given the rapid spread of information made possible by social media, a malicious attack on an organization can produce nearly instantaneous loss, as shown by the recent Domino's case on YouTube. The traditional approach to identifying risks (i.e, we are a chemical company, and chemicals sometimes spill) and predicting their trajectories through online communities needs to be expanded to include all possibilities through future research.

Transparency. Information transparency is the degree to which organizational actions and decisions are ascertainable and comprehensible by interested parties (Grunig & Hunt, 1984). Transparency is not just about information; it is a process whereby active participation in acquiring, distributing, and creating knowledge (Grunig & Huang, 2000) with stakeholders is essential to effective relationships. According to Reynolds, Galdo, and Sokler (2002) in their review of risk communication literature, empathy, caring, competence, commitment, and accountability contribute to trust. A fundamental characteristic of Web 2.0 is transparency.

According to Krimsky (2007):

> As the internet grew, access to information by communities, public interest groups and ordinary citizens has been unprecedented up to this point in human history. Expertise on risk became secondary to the significance of trust and validity in the source of information. Public skepticism is fed and nurtured by new open forms of internet communication. All risk websites are potentially equal in Cyberspace, constrained only by the skill of the web designer. It levels the playing field to a public that does not understand the hierarchy of expertise (meritocracy). (p. 160)

Prior Engagement. Every risk/crisis communication plan should have a section for communicating with stakeholders and working with the media. Social media can be used to both communicate directly with stakeholders and the media at the same time. More importantly, social media provides a built-in channel for stakeholders to communicate directly with the organizations. Incorporating social media into the plan ensures the tools will be analyzed and tested before a crisis and requires continual updating of the plan as social media evolves.

Individuals may have information that is crucial to the mitigation of the crisis, but if they do not trust the organization, or even know where to find it, that information will likely not be shared. In the midst of a crisis is not the time to suddenly try a hand at social media. To build partnerships and build trust, the discussion with publics should already be taking place. As Stephens and Malone (2009) identified in their research, if the organization wants to ensure that stakeholders go to their websites and new media technologies concerning new information, the organization must use the web prior to a crisis.

Vulnerable Populations. The environmental justice literature clearly demonstrates the differential impact of minority and socioeconomic challenged communities related to health, safety, and environmental issues, including the impact from disasters. According to Moore (2010), "The lack of a fully developed literature on the vulnerability of minority communities and institutions suggests that their critical role in the provisions of services to minority individuals is not well enough understood" (p. 5). Despite the uncertainty in this area, one thing from discussions is clear: new communication technology inequities plague vulnerable populations that can compromise how quickly and how much risk information they receive and generate during a crisis. Concerns include the need to examine message strategies to engage in dialogue to adopt risk-reducing behaviors, with consideration of the feasibility and ethicality of using social media with vulnerable populations who may or may not have the technical resources, access, or education to use social media during a risk or crisis event.

Conclusion

Taking full advantage of the opportunities provided by social media in pre-crisis communication planning and communication executions, while avoiding the potential pitfalls associated with these technologies, requires careful, thoughtful analysis long before any crisis emerges. Emergency managers should monitor potential issues while establishing crises communication plans that reside on both static and mobile devices. As issues are identified, research conducted on potential audiences should identify their trusted sources of information. Emergency managers and responders need to be proactive instead of reactive to take full advantage of the immediacy provided by mobile media.

The pre-crisis communications action plan should include policies specific to the use of social media in a risk or crisis event. A thorough risk assessment needs to be implemented to determine the challenges and opportunities of using social media in specific situations and the appropriate action steps that need to be taken to address any identified risks. It is crucial that everyone involved understands the nature of social media, how it is used, and expected behaviors by individuals using this technology in a crisis situation. Policies and measures for securing these devices should be in place. Security measures, new cyber attack protocols, and best practices in dealing with cyber terrorism must be addressed in relation to these new media technologies.

In general, emergency managers need to build a mobile communications and online community where stakeholders can engage with others and obtain contact information for media outlets and other crisis communication representatives. Empowering stakeholders in a crisis situation can work to everyone's advantage.

The CIO Council, in their September 2009 report entitled *Guidelines for Secure Use of Social Media by Federal Departments and Agencies*, concisely summarized the decision to use Web 2.0 media based on costs and benefits (risk based on a strong business case) as opposed to making decisions on a technology-based decision. For example, the decision to authorize access to social media websites is a risk-based business decision with inputs from all involved parties (Walls, 2007). Technology is not always or necessarily the answer, but sound new media emergency management integrated with risk, crisis, and issues management best practices can drive effective use of these evolving communication tools.

References

Aherton, J. (2009, October). Crisis planning in the digital age. *Frontlineonline.* International Public Relations Association. Retrieved from http://www.ipra.org/archivefrontlinedetail.asp?articleid=1400

American Red Cross. (2010). Safe and well list. Retrieved from https://disastersafe.redcross.org

———. (2010, August 9). Web users increasingly rely on social media to seek help in a disaster. Retrieved from http://www.redcross.org/portal/site/en/menuitem.94aae335470e233f6cf911df43181aa0/?vgnextoid=6bb5a96d0a94a210VgnVCM10000089f0870aRCR

Avery, E. J., & Kim, S. (2009). Anticipating or precipitating crisis? Health agencies may not be heeding best practice advice in avian flu press releases. *Journal of Public Relations Research, 21*(2), 187–197.

Baekel, T. (2008, June 23). The mobile internet revolution is here. Retrieved from http://www.baekdal.com/trends/mobile-internet-revolution

Berger, B. K., & Reber, B. H. (2006). *Gaining influence in public relations: The role of resistance in practice.* Mahwah, NJ: Lawrence Erlbaum.

Boyd, D. M, & Ellison, N. B. (2008). Social network sites: Definition, history, and scholarship. *Journal of Computer-Mediated Communication, 13,* 210–230.

Bridges, J. A. (2004). Corporate issues campaigns: Six theoretical approaches. *Communication Theory, 14*(1), 51–77.

Bulkely, K. (2010, June 18). Mobile technology takes centre stage in disaster relief. Retrieved from http://www.guardian.co.uk/activate/mobile-technology-disaster-relief

CIO Guidelines for secure use of social media by federal departments and agencies. (2009). Retrieved from http://www.cio.gov/Documents/Guidelines%5Ffor%5FSecure%5FUse%5FSocial%5FMedia%5Fv01%2D0%2Epdf

Cloudman, R., & Hallahan, K. (2006). Crisis communications preparedness among U.S. organizations: Activities and assessments by public relations practitioners. *Public Relations Review, 32*, 367–376.

Coombs, W. T. (2007). Protecting organization reputations during a crisis: The development and application of situational crisis communication theory. *Corporate Reputation Review*, *10*, 163–176.

———. (2008, April 2). Crisis communication and social media. Retrieved from http://www.instituteforpr.org/essential_knowledge/detail/crisis_communication_and_social_media/

Cordner, G., & Scarborough, K. (2010, January). Information sharing: Exploring the intersection of policing with national and military intelligence. *Homeland Security Affairs, 6*(1), 78–96.

Dake, K. (1992). Myths of nature: Culture and the social construction of risk. *Journal of Social Issues, 48*, 21–37.

Deragon, J. (2008). Leveraging social media for business purposes. White paper. Retrieved from http://www.slideshare.net/jderagon/Leveraging-Social-Media-for-Business

Dillard, J. P., Shen, L., & Vail, R. G. (2007). Does perceived message effectiveness cause persuasion or vice versa? 17 consistent answers. *Human Communication Research, 33*, 467–488.

Ellison, N. B., Steinfield, C., & Lampe, C. (2007). The benefits of Facebook "friends": Social capital and college students' use of online social network sites. *Journal of Computer-Mediated Communication*, *12*, 1143–1168.

Festinger, L., Pepitone, A., & Newcomb, T. (1952). Some consequences of de-individuation in a group. *Journal of Abnormal and Social Psychology, 47*, 382–389.

Fleisher, C. S. (2001). Emerging US public affairs practice: The 2000+ model. *Journal of Public Affairs, 1,* 44–52.

Freudenberg, N. (1984). *Not in our backyards! Community action for health and the environment.* New York: Monthly Review Press.

González-Herrero, A., & Smith, S. (2008). Crisis communications management on the web: How internet-based technologies are changing the way public relations professionals handle business crises. *Journal of Contingencies and Crisis Management, 16*(3), 143–153.

Gordon, J. (2009). The coming change in social media business applications. White paper *for Social Media Today*. Retrieved from http://www.socialscoop.biz/SMT_whitepaper_biz.pdf

Grunig, J. E. (2006). Furnish the edifice: Ongoing research in public relations as a strategic management function. *Journal of Public Relations Research, 18*(2), 151–176.

Grunig, J. E., & Huang, Y. H. (2000). From organizational effectiveness to relationship indicators: Antecedents of relationships, public relations strategies, and relationship outcomes. In J. A. Ledingham & S. D. Bruning (Eds.), *Public relations as relationship management* (pp. 23–54). Mahwah, NJ: Erlbaum.

Grunig, J. E., & Hunt, T. (1984). *Managing public relations*. New York: Holt, Reinhart & Winston.

Heath, R. L. (1996). Corporate environmental risk communication: Cases and practices along the Texas Gulf Coast. In B. R. Burleson (Ed.), *Communication yearbook 18* (pp. 255–277). Thousand Oaks, CA: Sage.

Heath, R. L. (2006). A rhetorical theory approach to issues management. In C. H. Botan & V. Hazleton (Eds.), *Public relations theory II* (pp. 63–99). Mahwah, NJ: Lawrence Erlbaum.

Heath, R. L., Bradshaw, J., & Lee, J. (2002). Community relationship building: Local leadership in the risk communication infrastructure. *Journal of Public Relations Research, 14,* 317–353.

Heath, R. L., Lee, J., & Ni, L. (2009). Crisis and risk approaches to emergency management planning and communication: The role of similarity and sensitivity. *Journal of Public Relations Research, 21*(2), 123–141.

Heath, R. L., & Nathan, K. (1991). Public relations' role in risk communication: Information, rhetoric and power. *Public Relations Quarterly, 35*(4), 15–22.

Heath, R. L., & Palenchar, M. J. (2009). *Strategic issues management: Organizations and public policy challenges* (2nd ed.). Thousand Oaks, CA: Los Angeles.

Heitmann, H., & Lott, B. (2008). Protecting corporate reputation in an era of instant transparency. In D. Pantaleo & N. Pal (Eds.), *From strategy to execution: Turning accelerated global change into opportunity* (pp. 237–257). Berlin: Springer.

Hinduja, S., & Patchin, J. W. (2008). Personal information of adolescents on the Internet: A quantitative content analysis of MySpace. *Journal of Adolescence, 31,* 125–146.

Ingram, M. (2010, April 12). Mary Meeker: Mobile internet will soon overtake fixed internet. Retrieved from http://gigaom.com/2010/04/12/mary-meeker-mobile-internet-will-soon-overtake-fixed-internet/

Jaques, T. (2010). Embedding issue management as a strategic element of crisis prevention. *Disaster Prevention and Management, 19*(4), 469–482.

Krämer, N. C., & Winter, S. (2008). Impression management 2.0: The relationship of self-esteem, extraversion, self-efficacy, and self-presentation within social networking sites. *Journal of Media Psychology, 20(3),* 106–116.

Krimsky, S. (2007). Risk communication in the internet age: The rise of disorganized skepticism. *Environmental Hazards, 7,* 157–164.

Lariscy, R. W., Avery, E. J., Sweetser, K. D., & Howes, P. (2009). An examination of the role of online social media in journalists' source mix. *Public Relations Review*, DOI:10.1016/j.pubrev.2009.05.008.

Mangold, W. G., & Faulds, D. J. (2009). Social media: The new hybrid element of the promotion mix. *Business Horizons, 52*, 357–365.

Marken, G. A. (2007). Social media....The hunted can become the hunter. *Public Relations Quarterly*, *52*(4), 9–12.

Mazer, J. P., Murphy, R. E., & Simonds, C. J. (2007). I'll see you on Facebook: The effects of computer-mediated teacher self-disclosure on student motivation, active learning, and classroom climate. *Communication Education, 56*, 1–17.

McKay, L. (2009, June). Everything's social (now). *Customer Relationship Management*, 24–28.

Mitchell, R. K., Agle, B. R., & Wood, D. J. (1997). Toward a theory of stakeholder identification and salience: Defining the principle of who and what really counts. *Academy of Management Review*, *22*(4), 853–886.

Moore, T. (2010, May). *Institutional barriers to resilience in minority communities.* Institute for Homeland Security Solutions.

Musil, S. (2008, October 26). U.S. Army warns of twittering terrorists. CNET News. Retrieved from http://news.cnet.com/8301-1009_3-10075487-83.html

Nichols, R. (2010, June 7). Emergency text messaging signals evolution in public safety communication. Retrieved from http://www.emergencymgmt.com/safety/Emergency-Text-Messaging-Public-Safety-Communication.html

O'Dell, J. (2010, April). New study shows the mobile web will rule by 2015. Retrieved from http://mashable.com/2010/04/13/mobile-web-stats/

Palen, L., Vieweg, S. J. Sutton, Liu, S. B., & Hughes, A. (2007). Crisis informatics: Studying crisis in a networked world. *Third International Conference on Social Science*, Ann Arbor, Michigan, October 7–9, 2007.

Palenchar, M. J., & Heath, R. L. (2002). Another part of the risk communication model: Analysis of communication processes and message content. *Journal of Public Relations Research, 13,* 127–158.

Palese, M., & Crane, T. Y. (2002). Building an integrated issue management process as a source of sustainable competitive advantage. *Journal of Public Affairs, 2,* 284–293.

Pavlik, J. (2008, June 4). Mapping the consequences of technology on public relations. Retrieved from http://www.instituteforpr.org/essential_knowledge/detail/mapping_the_consequences_of_technology_on_public_relations

Pearson, C. M., & Clair, J. A. (1998). Reframing crisis management. *Academy of Management Review, 23*(1), 59–76.

Pew Research Center. (2010). *Government online: The internet gives citizens new paths to government services and information.* Washington, DC: Pew Internet & American Life Project.

Prentice, S., & Huffman, E. (2008, March). Social media's new role in emergency management. *Idaho National Laboratory.* Proceedings from *Emergency Management and Robotics for Hazardous Environments*. Albuquerque, New Mexico. March 9–12, 2008.

Rand, P. M., & Rodriguez, G. (2007). Relating to the public: The evolving role of public relations in the age of media. *The Council of Public Relations Firms*. White paper. Retrieved from 20White%20Paper%20FINAL.pdf

Renn, O., Burns, W. J., Kasperson, J. X., Kasperson, R. E., & Slovic, P. (1992). The social amplification of risk: Theoretical foundations and empirical applications. *Journal of Social Issues, 48*, 137–160.

Reynolds, B., Galdo, J., & Sokler, L. (2002). *Crisis and emergency risk communication.* Atlanta, GA: Centers for Disease Control and Prevention.

Reynolds, B., & Seeger, M. W. (2005). Crisis and emergency risk communication as an integrative model. *Journal of Health Communication, 10,* 43–55.

Roper, J. (2002). Government, corporate or social power? The Internet as a tool in the struggle for dominance in public policy. *Journal of Public Affairs, 2,* 113–124.

———. (2005). Symmetrical communication: Excellent public relations or a strategy of hegemony? *Journal of Public Relations Review, 17,* 69–86.

Roux-Dufort, C. (2007, June). Is crisis management (only) a management of exceptions? *Journal of Contingencies and Crisis Management, 15*(2), 105–114.

Semple, E. (2009). Update your crisis communication plan with social media. *Strategic Communication Management, 13*(5), 7.

Smith, A. (2010, April). *Government online: The internet gives citizens new paths to government services and information.* Washington, D.C.: Pew Internet & American Life Project.

Solis, B. (2008a). *The essential guide to social media*. Retrieved from http://www.briansolis.com

———. (2008b). *Reinventing crisis communications for the social web*. Retrieved from http://www.briansolis.com

Spencer, T. (2004). Of legitimacy, legality and public affairs. *Journal of Public Affairs, 4,* 205–209.

Steinfield, C., Ellison, N. B., & Lampe, C. (2008). Social capital, self-esteem, and use of online social network sites: A longitudinal analysis. *Journal of Applied Developmental Psychology, 29,* 434–445.

Stephens, K. K., & Malone, P. C. (2009). If the organizations won't give us information…: The use of multiple new media for crisis technical translation and dialogue. *Journal of Public Relations Research, 21(*2), 229–239.

Stephenson, D., & Bonabeau, E. (2007, February). Expecting the unexpected: The need for a networked terrorism and disaster response strategy. *Homeland Security Affairs, 3*(1). Retrieved from http://www.hsaj.org

Sutton, J. N. (2010). *Twittering Tennessee: Distributed networks and collaboration following a technological disaster.* Paper presented at the 7th International ISCRAM Conference, Seattle, WA.

Terdiman, D. (2009, January 15). Photo of Hudson River plane crash downs TwitPic. *CNET.* Retrieved from http://www.news.cnet.com

Traynor, P. (2008, September). *Characterizing the limitations of third-party EAS over cellular text messaging services.* Atlanta: Georgia Institute of Technology.

Vanderford, M. L., Nastoff, T., Telfer, J. L., & Bonzo, S. E. (2007). Emergency communication challenges in response to Hurricane Katrina: Lessons from the Centers for Disease Control and Prevention. *Journal of Applied Communication Research, 35*(1), 9–25.

Vaughan, E., & Seifert, M. (1992). Variability in the framing of risk issues. *Journal of Social Issues, 48*, 119–135.

Vieweg, S., Palen, L., Liu, S., Hughes, A., & Sutton, S. (2008). Collective intelligence in disaster: An examination of the phenomenon in the aftermath of the 2007 Virginia Tech shootings. Proceedings of the Information Systems for Crisis Response and Management Conference. Washington, D.C.

Walls, A. (2007). *Corporate use of social networks requires multilayered security control.* Gartner Research.

Walther, J. B., Van Der Heide, B., Kim, S. Y., Westerman, D., & Tong, S. T. (2008). The role of friends' appearance and behavior on evaluations on Facebook: Are we known by the company we keep? *Human Communication Research, 34*, 28–49.

———. (2007). Selective self-presentation in computer-mediated communication: Hyperpersonal dimensions of technology, language, and cognition. *Computers in Human Behavior, 23*, 2538–2557.

Waters, R. D., Burnett, E., Lamm, A., & Lucas, J. (2009). Engaging stakeholders through social networking: How nonprofit organizations are using Facebook. *Public Relations Review, 35*, 102–106.

Wright, D. K., & Hinson, D. H. (Spring 2008). How blogs and social media are changing public relations and the way it is practiced. *Public Relations Journal, 2*(2), 1–21.

———. (Spring 2009). An updated look at the impact of social media on public relations practice. *Public Relations Journal, 3*(2), 2–27.

Yang, S. & Kang, M. (2009). Measuring blog engagement: Testing a four-dimensional scale. *Public Relations Review, 35*(3), 323–324.

Zhao, S., Grasmuck, S., & Martin, J. (2008). Identity construction on Facebook: Digital empowerment in anchored relationships. *Computers in Human Behavior, 24,*1816–1836.

Chapter 9

Efficacy of Crisis Management Plans: Toward a Revised Synthesis Model in Managing Crises

Luana Sok-Yee Low, Venetia Tzi-Wei Chung & Augustine Pang

The authors present insights into the essential use of crisis management plans (CMPs) for crisis planning. Couched within the framework of Horsley & Barker's (2002) synthesis model of crisis, these authors conducted interviews with eight public relations professionals. Analysis of the interview transcripts revealed that CMPs help foster organizational vigilance, a healthy organizational culture, and proactive communication with stakeholders. Based on their results, the authors proposed additions to the synthesis model.

Introduction

This book chapter explores the importance of the crisis management plan (CMP) and its applicability in a crisis. Using Horsley & Barker's (2002) Synthesis Model as their theoretical lens, the authors sought to uncover other components in the cyclical model with the aim of developing a more holistic crisis model. This study found that the CMP played the critical role of providing a platform to engage organizations to develop a systematic program in stakeholder mapping; risk and issues identification; delineating the roles and responsibilities of the crisis management teams; and developing contact lists for different stakeholders. The study also found two trigger points that call for the activation of the CMP: the first being the gravity of the situation and second, when industry partners encounter crisis. With these insights, we propose to expand Horsley and Barker's (2002) Synthesis Model by positing three more stages, namely, propagating a healthy organizational culture; practicing proactive external communications; and the activation of the CMP, which will depend on predefined trigger points that will signal organizations when to activate their crisis plans.

Overview

Most studies on crisis management are premised on how organizations can and should develop a crisis management plan (CMP) to prepare, respond, and recover from crises. Indeed, the lynchpin that determines the success or failure of a crisis campaign appears to rest on an organization's crisis preparedness, epitomized by the presence of a crisis plan (Cloudman & Hallahan, 2006; Coombs, 2000). For some organizations, the CMP serves as insurance that will help them negotiate through crises. Studies on public-listed, transnational organizations, however, found that many do not have a CMP in place (Burnett, 1998; Fink, 1986; Pinsdorf, 1995; Tiller, 1994). Beyond the CMP, studies have found that other factors play equally important roles in determining how well organizations deal with crises. Marra (2004), for instance, argued that the autonomy of public relations (PR) practitioners and an empowering organizational culture were critical factors in managing crises.

Given the limitations of the CMP as a working document, it would be erroneous to believe it would meet all potential challenges (Horsley & Barker, 2002; Pang, Cropp, & Cameron, 2006). Besides, Coombs (2000) argued that little is known about what crisis managers do before and during a crisis and where the CMP fits into overall crisis planning. There is a need to revisit how organizations use the CMP and to examine what other impact the execution of crisis plans has. Using Horsley and Barker's (2002) Synthesis Model as its theoretical lens, this study seeks to expand the cyclical model to develop a more holistic model that will aid in effective crisis planning. Consisting of five interconnected stages, the Synthesis Model is premised on the idea that all crisis communication activities revolve around the internal and external environments of an organization. The five interconnected stages are: (1) ongoing public relations efforts, (2) identification of and preparation for potential crises, (3) internal training and rehearsal, (4) the crisis event, and finally (5) evaluation and revision of public relations efforts. While this model has provided a template for public sector communication, Liu and Horsley (2007), however, argued that this model was "limited" by its focus on government crisis communication (p. 382).

Data to examine the questions in this study come from in-depth interviews with an exploratory group of eight public relations consultants with extensive experience in crisis communication and management. Though the interviews were conducted in Singapore, described as a developed economy and a key financial center in Asia with a cosmopolitan and multiethnic society (Chay-Nemeth, 2003; Lim, Goh, & Sriramesh, 2005), the interviewees represented a diverse range of industries, came from different countries, and possessed valuable international experience. Their views would thus represent international rather than local perspectives. The interviewees were asked first how the CMP enhances crisis planning and second what other factors would trigger the activation of the CMP.

This study aims to add to discussions on components in models of crisis management (see Barton, 1994; Borda & Mackey-Kallis, 2004; Mitroff & Anagnos, 2001; Olaniran & Williams, 2001; Pearson & Mitroff, 1993) and answers the call by the authors to further test the applicability of the model.

Second, this model was conceived to examine crisis communication in the public sector. However, it is instructive for the private sector as well. This exploratory study aims to examine applicability of the model to the private sector and to build on the model.

Third, with the Synthesis Model as an initial template, it is hoped that this would open up possibilities for future studies to explore how it can be made more realistic. Pang, Cropp and Cameron (2006) argued that studies like this, grounded in the practitioners' world, provide important contexts on how theory and practice can integrate.

Literature Review

Recurring Theme in Crisis-preparedness: Control

Seeger, Sellnow, and Ulmer (1998) defined crisis as a negative event that is unexpected and non-routine. Pinsdorf (1987) argued for the inevitability of crises affecting organizations. Crises are "no longer a matter of if, but when; no longer an exception, but the expected, even the inevitable" (p. 37). Marra (2004) argued that crises are inevitable, and organizations need to plan for the possibilities of different types of crises that they may encounter. This is often done through strategic planning, which involves projecting the conditions of a crisis and identifying the resources, structures, and strategies necessary to resolve the crisis with as little disruption, cost, and harm as possible (Seeger, Sellnow, & Ulmer, 2003). Given that most studies on crisis communication are focused on developing multifaceted plans to prepare, respond, and recover from crises (Pang, Cropp, & Cameron, 2006), McConnell and Drennan (2006) argued that despite the difficulties in translating the ideals of crisis preparation into practice, the post-9/11 period made crisis planning more pertinent, heightened, relevant, and important. Challenges to crisis preparation include the inefficient allocation of resources to prepare for low-probability crisis events, the unpredictability of crisis often rendering crisis plans ineffectual, the fragmentation of networks within organizations making robust planning difficult, and the costliness of executing training and drills.

Organizations consistently fail to embark on crisis planning. Studies in the 1980s found that among all organizations, less than 50 percent had any crisis plan (Fink, 1986). In the 1990s, among the Fortune 1,000 Industrial and Fortune 500 service firms, only 53 percent had crisis plans (Pinsdorf, 1995). Among the Fortune 1,000 organizations, less than 60 percent had crisis plans (Tiller, 1994). Among the largest profit-making organizations in the U.S., up to 70 percent of them had not made any crisis plans (Burnett, 1998). In the 2000s, only larger companies were found to have crisis plans (Cloudman & Hallahan, 2006). Several reasons are proffered for this apparent malaise. One, organizations are under the illusion that a crisis would not happen to them. Second, they do not have the expertise to embark on crisis planning. Some may seek the help of public relations consultancies (Pang & Yeo, 2010), but these often involve enormous costs. Third, there may be a lack of impetus to embark on crisis planning when there are more important daily operational concerns. Fourth, the opportunity costs of crisis planning are deferred in favor of more immediate concerns.

Traditionally, an organization's crisis-preparedness is epitomized by a crisis plan (Coombs, 2000) that is set against the crisis life cycle. For instance, Fink's (1986) four-stage model comprised the prodromal, acute crisis, chronic crisis, and crisis resolution stages. The first stage is the pre-crisis point where hints of a potential crisis begin to emerge; the acute stage, also known as the turning point of the crisis, is when the crisis occurs and triggers subsequent external reactions. The third stage, chronic crisis, is when outstanding issues from the immediate crisis are addressed. Also known as the clean-up phase, organizations engage in a self-reflective process and take corrective actions when necessary. Finally, the crisis resolution stage is a forward-looking phase where organizations pick out learning points from the previous crisis to plan and pre-empt for future crises. Crisis planning is necessary as it helps organizations better control crisis situations.

Mitroff & Anagnos (2001) stressed the importance of pre-crisis planning. They introduced several guiding principles: (1) identifying crisis types and their accompanying risks; (2) understanding the importance of mechanisms for anticipating, sensing, reacting, containing, and redesigning effective organizational procedures before, during, and after a crisis; and (3) understanding the systems that govern both internal and external operations, such as the culture of the workplace and identifying key stakeholders. Cultivating positive relationships with stakeholders can also help in crisis mitigation. If planning, as argued, is of paramount importance, one symbol of crisis preparedness is the conceptualization of the crisis management plan.

Crisis Management Plans: Symbols of Crisis Planning

Crisis planning is considered to be the "most critical" (Seeger, Sellnow and Ulmer, 2003, p. 164) strategic activity an organization could undertake. Gonzalez-Herrero and Pratt (1996) argued it is the "bedrock" (p. 96) of crisis management. Pang, Cropp, & Cameron, (2006) argued it involved two overarching components: first, identifying possible conditions/scenarios of crises; second, allocating resources, capacities, and structures to deal with these conditions. The following components form the core of a crisis plan, argued Pang, Cropp, and Cameron (2006). The authors recommend identifying general and specific vulnerabilities (Lerbinger, 1997), key stakeholders (Gonzelez-Herraro & Pratt, 1996), key team members (King, 2002), and communication channels (Fearn-Banks, 2007). In his CMP framework, Coombs (2007) included other components—such as a crisis management team contact sheet, crisis risk assessment, crisis communication team communication strategy worksheet, and crisis control center description—that bring together background information, decision-makers, and the protocols an organization should observe when a crisis occurs.

Given that the goals of crisis planning were to reduce uncertainty, decision time, response, and recovery, and the purpose of conceptualizing the CMP is championed as a strategic tool to help enhance an organization's reputation, image and credibility (Lerbinger, 1997; Seeger, Sellnow, & Ulmer, 2003), one strong component in the plan must be the communication dimension in crisis management. This is because, as Marra (1998) argued, when a crisis occurs, a barrage of questions would be aimed at the organization as stakeholders want to know what happened, how it happened, when, why, and how the crisis will affect them. This emphasizes the importance of effective communication in a crisis plan.

A working and constantly updated CMP helps organizations respond to crisis in a timely and effective manner (Seeger, Sellnow, & Ulmer, 2003). The CMP is at best a reference tool, not a regimented manual (Coombs,

2007). In some organizations, the CMP offers reassurance to external and internal publics. A CMP helps to limit the harm, duration, severity, and intensity of the crisis (Seeger, Sellnow, & Ulmer, 2003). In the face of severe and unexpected threats where there is a need for urgency in decision-making (McConnell & Drennan, 2006), the CMP is a powerful tool that facilitates this process. Gonzalez-Herrero and Pratt's (1996) Integrated Symmetrical Model for Crisis Communication Management argued that crisis planning could mitigate a crisis, which in turn could help build positive relationships. Feedback from stakeholders could further act as an impetus to improve on the plan.

However, it must be recognized that the CMP is not an exhaustive document to be religiously followed. Seeger, Sellnow, & Ulmer (2003) highlighted that although a crisis plan may be useful in reducing time pressure through the predetermination of some decisions and procedures, this does not alleviate the risks that organizations face. Moreover, the diversity of issues the organization faces every day makes it difficult to pre-empt any crisis it might face. Ultimately, the CMP should not be seen as an end in itself but as a means to an end. As Ulmer (2001) argued, the end of every crisis is the beginning of the preparation step for the next one. Thus, as a secondary measure, the CMP can also be used to evaluate the handling of a crisis so as to further prepare the organization for future crises.

Limitations of the CMP: Usability Questioned

Previous studies have shown that the CMP may not be the best tool to have in mitigating a crisis. Even with a CMP in place, there could be hindrances to its usability. Coombs (2007) argued that many plans have been described as unnecessarily long and burdensome. For plans to be useable, they must be reader-friendly and followed up by action.

The CMP has also been criticized for its reactive nature, which is the result of much focus being placed on tackling the last crisis rather than planning for future threats (McConnell & Drennan, 2006). This causes the plan to be specific to a particular crisis type (Seeger, Sellnow, & Ulmer, 2003). More importantly, the formation of CMPs is mostly based on past experiences, which automatically limits the underlying potential of the plan. This process of sensemaking (Weick, 2001) delineates invisible boundaries, which are limited by experience. The CMP can also be mired in details, thus losing its macroscopic functions (Marra, 2004). Even more problematic is the idea that organizations may be under the false impression that they are immune to crises because they have a CMP. Given the limitations, it would be erroneous to believe in a "one size fits all" (Horsley & Barker, 2002, p. 428) plan that covers all potential challenges. Crisis plans therefore have to be sensitive to the fluid conditions of risks and threats and, more importantly, be able to be applied to the different paradigms that these organizations operate in.

Thus, even the most synergized and comprehensive crisis plan is only as good as it looks on paper, as it does not reflect the real threats and dilemmas that crises in reality present. As McConnell and Drennan (2006) argued, "to conceive of 'ideal' preparedness is a paradox. How can we prepare for something if we do not even know what we are preparing for?" (p. 57). The usability of the CMP is further hindered by a multitude of challenges that organizations face. First, the development of a CMP requires investment of time and money, as pre-planning requires the cooperation of fragmented and often-conflicting interests (McConnell & Drennan, 2006). Furthermore, the bureaucracy within large organizations makes it difficult for horizontal and vertical integration (McConnell & Drennan, 2006), stifling opportunities to develop a CMP. A typical profit-driven organization would channel more resources toward the frontline businesses of the organization rather than worry about the possibilities of crises (Pang, Cropp & Cameron, 2006). As such, many organizations would be less inclined to invest resources into something they perceive to have low market value.

As discussed above, another barrier to the development of the CMP lies with the negative inclinations the dominant coalition has towards investing in crisis management. Developing a CMP often affects the functional arms of the organization, as it involves necessary change. Thus, it creates a tension between reformists and the conservatives (McConnell & Drennan, 2006) in the organization. Due to the limited payoff of contingency planning, it can be a hard-sell to convince the dominant coalition to allocate even minimal amounts of resources to the creation of a CMP.

Seeger, Sellnow and Ulmer (2003) argued the reluctance in organizations to discuss crisis and the risks they face as yet another possible difficulty in initiating a CMP. Many organizations are afraid that disclosing information sensitive to their businesses could invite suspicion from stakeholders. They fear openly discussing risks and threats would influence the stakeholders' confidence in the organization. Thus, their willingness to undertake planning is undermined by the very fact that they do not acknowledge the possibility of crises.

Beyond the CMP: Other Factors at Work

Scholars have proposed other factors involved in crisis planning. Penrose (2000) explained that the perception of a crisis as an opportunity or a threat affects an organization's willingness to engage in crisis management activities, which in turn affect the outcome of a crisis. Marra (2004) argued the importance of autonomy accorded to communication professionals and an empowering organizational culture.

As argued above, it would be foolhardy for organizations to simply rely on the CMP to resolve the extraordinary and unpredictable circumstances of a crisis. Possessing a crisis communication plan does not necessarily translate into an organization's ability to implement it (Coombs, 2007). Thus, considerable room for individual autonomy in responding to such situations is required, as the amount of autonomy can impact on how well an organization communicates during a crisis. A CMP is also of limited use if it does not align with an organization's philosophies, values, attitudes, assumptions, and norms. Thus, the successful adoption of the CMP depends on how the organization reflects organizational philosophy and culture on the plan. An open organizational culture, for instance, may lead to a comparatively more open communication policy (Penrose, 2000).

Yet, another factor that affects the outcome of a crisis is the relationships that organizations share with their stakeholders. This is based on the stakeholder theory (Freeman, 1984; Freeman & Gilbert, 1987) that organizations need to look beyond just their shareholders. In a crisis, stakeholders want to know how a crisis can affect them. The building of mutually beneficial relationships (Heath, 1997) would help to maintain communication channels, which are more crucial to a successful handling of a crisis than its CMP. Ultimately, positive relations would help to build reservoirs of good will and shared understanding (Ulmer, 2001). One example of the lack of communication to stakeholders is the case of the *Challenger* space shuttle in 1986, whereby a media blackout from NASA raised much negative conjectures through the handling of the situation. As Marra (2004) highlighted, in the absence of information, misinformation becomes news, and this could compromise the image of the organization, possibly creating negative repercussions. Despite the fact that NASA issued a statement clarifying the facts after the media blackout, it was too little too late in assuaging public fears. Thus, effective communication with stakeholders should be an organization's top priority during a crisis. Additionally, some scholars advocate the importance of communication even before a crisis. This is clearly defined in the case of Malden Mills. The textile organization suffered an industrial accident that threatened the jobs of

3,000 employees. The success of Malden Mills' communication was attributed to the positive pre-crisis relationships established among its stakeholders founded on understanding and trust (Ulmer, 2001). The unhindered flow of information to its stakeholders during the industrial fire that engulfed the factory rendering the curtailment of operations for weeks helped to allay the fears of the thousands of employees whose livelihoods in small town Lawrence, Massachusetts, depended on the factory.

Related to stakeholder theory is the idea that organizations who seek to understand the emotional upheavals faced by stakeholders during crises would be in a better position to reach out to them in more effective ways. The integrated crisis mapping model posits four dominant emotions stakeholders are likely to feel during crisis: anger, sadness, fright, and anxiety (Jin, Pang, & Cameron, 2007). However, anxiety is the initial, dominant, and default emotion stakeholders always feel during a crisis (Jin, Pang, & Cameron, 2009). Their concern is about how the situation will resolve, the future, and what the organization would do for them. Understanding stakeholders' emotions is argued to be the new frontier in crisis research (Jin & Pang, 2010), and organizations would do well to craft a mix of both defensive and accommodative strategies that reach out to them intellectually, physically, and materially (Pang, Jin, & Cameron, 2010a).

Synthesis Model of Crisis Communication

Given the many factors at play beyond the mere conceptualization of the CMP, the challenge is to synthesize these factors and develop as encompassing an approach as possible in managing crisis. Recognizing that no two crises are identical and that no single crisis communications plan will solve every problem, Horsley & Barker (2002) have proposed a synthesis model. The five interconnected stages of the Synthesis Model of Crisis Communications are:

Stage 1: Ongoing Public Relations Efforts. The most effective crisis communication plans begin well before the crisis event. Conscientious image building and continuous attention to public and media relations in good times ensure that communication systems are in place when they are needed most. A good rapport with the media helps organizations get their messages out during a crisis, and a good relationship with customers and stakeholders enhances the credibility of those messages.

Stage 2: Identification of and Preparation for Potential Crises. Good communicators identify potential problems and prepare for them. All crises cannot possibly be predicted, but good plans can be adapted to the situation by trained personnel.

Stage 3: Internal Training and Rehearsal. A written crisis communication plan, enhanced by training and practice drills, makes sure that all members of a crisis team know their roles and are prepared to handle their assigned task.

Stage 4: The Crisis Event. When a crisis occurs, prompt, concerted communication helps manage rumours and speculations; using the media to diffuse information is a key element in maintaining good public relations. Resolving the crisis in an ethical and humane manner is crucial.

Stage 5: Evaluation and Revision of Public Relations Efforts. After a crisis has passed, public relations efforts must be evaluated and revised in hindsight. An organization's leaders need to demonstrate that they have recovered from the crisis, have made changes to prevent it from happening again, and are prepared to deal with the next crisis that comes along. An organization can create additional media opportunities by providing follow-up stories and progress reports and thus handle a negative situation in a positive way. Good planning for public relations efforts immediately following a crisis can set the stage for future public relations efforts.

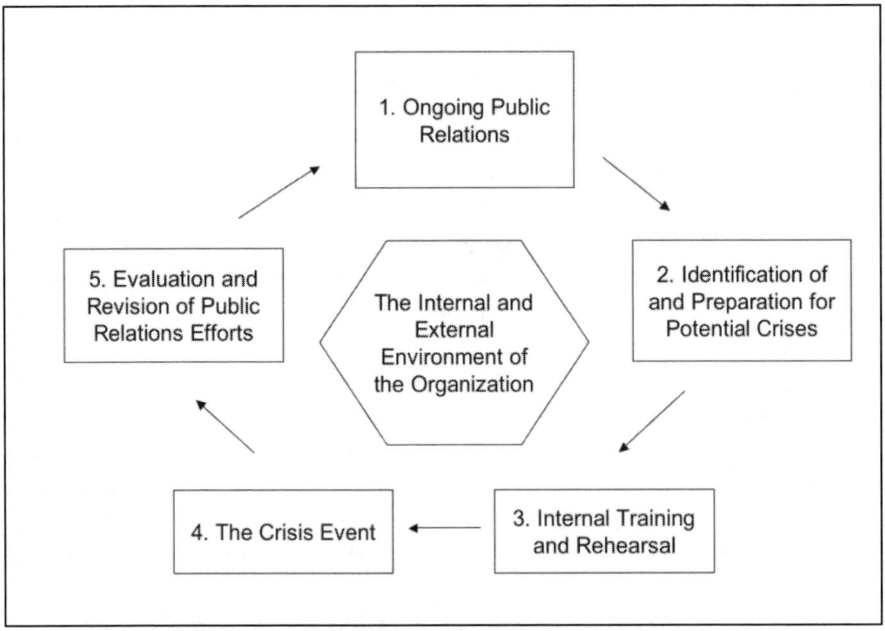

Figure 1: Synthesis Model of Crisis Communication

Horsley and Barker (2002) emphasized that throughout the process, both communication and management should be an ongoing effort on the organization's part. Framing the model in a cyclical manner highlights that crisis communication does not stop when the current crisis ends. They also further highlighted the importance of planning by advocating that organizations should be proactive in terms of scanning for issues and potential crises, rather than reacting only when a crisis occurs. Though this model was conceived with the public sector in mind, it is also instructive for the private sector.

Drawback of the Model

In every organization, there are many groups of stakeholders the organization needs to be accountable to, with or without a crisis. These groups are an important audience, as they often are most directly impacted when a crisis occurs. By positing ongoing public relations as the first stage of the Synthesis Model, Horsley and Barker (2002) recognized the importance of actively engaging their external stakeholders on a regular basis. Stakeholder mapping is essential in a crisis communication strategy as stakeholders, directly or indirectly, are the ones who form enduring perceptions of the organization when a crisis strikes. Therefore, building and maintaining positive relationships with them would be tantamount at any stage of a crisis. It would also make for easier conveying of messages during a crisis as stakeholders who are already in regular communication with an organization will tend to be more receptive and understanding due to the goodwill developed over time.

At the same time, there is also a need for organizations to understand that not all stakeholders will react in the same desired manner to the public relations efforts. Some may be receptive; others may not. Therefore, organizations need to understand and learn how to manage various groups of stakeholders who might have different reactions to a crisis. This ensures different strategies are used to manage the emotions of the stakeholders. Another aspect that is not sufficiently explored is the need to create a positive culture within the organization. While the model's five stages revolve around the internal and external environment of the organization, Horsley and Barker (2002) did not explicitly highlight the importance of ensuring that employees feel a sense of belonging to the organization. This, in fact, is one of the most important components of crisis management, since employees are the best spokespersons for the organization, and a healthy culture would propagate effective communication within the organization.

When employees have a sense of belonging, there is a greater tendency for them to assist in mitigating a crisis by communication of positive messages to the public.

Finally, most of the recommended steps in the Synthesis Model focus on internal efforts. The model does not take into consideration the impact of external factors and voices that might affect the strategies needed to be employed. Ongoing public relations may help to fortify a positive image for the organization, but in an open environment, there are also negative voices that organizations need to flag. Efficiently identifying the external forces will help to better shape the organization's crisis plan as the focus might shift based on the sentiments reflected by the dominant voices in the crisis.

Having discussed the various benefits and shortcomings of the Synthesis Model, it does present a reasonably comprehensive view of crisis management. At the same time, the model effectively links the various stages into a seamless flow. This makes it an easy-to-follow model, which organizations can adopt in their crisis planning and can springboard organizations to begin developing crisis plans. However, further improvements can be made to enhance the model's rigor and applicability. Thus, based on the literature, we seek to understand the role and usefulness of the CMP and different benchmarks in mitigating a crisis and how these different factors can culminate the ideal crisis plan for organizations. The research questions are: *How could the CMP enhance crisis planning?* and *Besides the CMP, what other factors are involved in crisis planning?*

Method

To develop a firmer grasp of the topic, the qualitative approach was used to discover how organizations view the CMP as a tool in mitigating a crisis. It allowed both the researcher and participant to delve into the complexity of the topic and capture the essence of the study.

Data Collection

In-depth Interviews. Given the sensitive nature of the topic, in-depth interviews were used because they create familiarity (Knapp, Miller & Fudge, 1994) with the interviewee. The questions were largely derived from the research questions, utilizing a flexible framework of questions, which allowed the interview to progress spontaneously. The confidentiality in an interview conveyed a sense of professionalism on the premise that the releasing of such organization-sensitive information could put respondents on their guard.

As this study is exploratory, elite interviews were conducted with an initial pool of eight seasoned practitioners. They were senior manager to managing director levels in public relations consultancies and possessed experience ranging between six and more than 20 years. Though they are now working in consultancy roles, many had in-house experience. All had international experience in corporate and government sectors. At the time of the interviews, they were principals or lead counsels in crisis management practices in their respective organizations. To ensure validity of the study, respondents were critically chosen based on their credentials. Here are the profiles of the interviewees:

Respondent A is currently an Account Manager at one of the top-three leading public relations consultancies in Singapore. Specializing in the high-risk healthcare sector, she has been in the media industry for more than eight years. Her portfolio in crisis management includes the handling of the SARS crisis that hit Hong Kong and Singapore, as well as her involvement with a voluntary product recall in the optometry industry. With her experience in the handling of many crises in the healthcare sector, the respondent's previous experience as a journalist provides great insight to the intricacies of the relationship between journalism and PR.

Respondent B is currently leading the corporate communications department at a Singapore government ministry. Previously the lead in crisis and issues management practice at a public relations firm that has won the Gold Award three times in Singapore, she garnered much attention as a news anchor on Asia's leading news channel. With almost 10 years of experience in the media industry, she was the client lead for a crisis in the medical sector in 2007, whereby a widely prescribed diabetes drug was reported to raise the risk of cardiovascular-related death and heart attack.

Respondent C is currently a senior manager at a leading global public relations firm, touted to be one of the largest agencies in the world. At the Singapore office, she is one of the key personnel directly assisting the Market Leader. With six years of experience in the industry, she has serviced many accounts in a wide range of sectors—from technology to healthcare, corporate and government. Her current specialization is in the government sector, where she works with the government ministry on several nationwide campaigns to promote society's acceptance of law offenders.

Respondent D is the founder and managing director of a local PR consultancy, which was set up in 1991. With a wide range of accounts from the technology to the food and beverage sectors, the interviewee has more than 20 years of experience in public relations and is often the organization's lead in the management of crisis issues. One of his major involvements included a major client in the food sector, who was struck by a food poisoning scare in December 2007, where the client had to revamp and rebuild its entire image to regain the trust of the public.

Respondent E is the principal consultant for a local PR consultancy. Accumulating more than 20 years of experience, he was a former President of the Institute of Public Relations of Singapore, as well as the ex-chair of the ASEAN Federation of Public Relations Organizations. He brings a wealth of knowledge extending to the food and beverage sector, where he handled the account for a top-selling soft drink as the Group Corporate Public Relations Manager for a global conglomerate.

Respondent F is currently a director at a leading global public relations firm, who has moved from the firm's office in the Netherlands to Japan and is currently based in Singapore. He currently heads the teams for major clients in the Corporate Government sector. In his nine years of experience in public relations, he has acquired extensive experience in crisis communications with major clients in the Singapore Government Ministry. Back in the Netherlands, he specialized in the healthcare sector, where he was highly involved with product recalls in the medical industry. His experiences both in Europe as well as in Asia Pacific and Japan bring an international perspective to crisis management.

Respondent G is currently a freelance public relations practitioner, who has been in the industry for more than 15 years. Her broad range of experience in public relations includes major clients from the technology and the finance sectors.

Respondent H is currently a director at a global public relations firm who has accumulated more than 20 years of experience in the technology sector. She heads the team for the organization's major technology client, in Singapore as well as in the Asia Pacific. Prior to her current appointment, she led the Corporate Communications team in the same technology firm that she is now servicing.

The interviews were conducted between September and October 2008 in Singapore. The questions asked included how organizations prepare for crisis, the role of the CMP in crisis preparation, and activation of crisis plans before exploring other factors involved in crisis planning.

Data Analysis

Thematic Analysis. To analyze the responses, interview transcripts were encoded based on a thematic classification, whereby the qualitative data were coded via a list of themes that were broadly related to one another (Boyatzis, 1998). Informed by the literature and the model, broad themes were developed through categorizing the raw information and identifying similar patterns among the information (Campbell & Sato, 2009). The information was then coded to the corresponding themes identified, while all the responses that fit the specific patterns were placed under the corresponding themes (Aronson, 1994). To ensure validity, the data were consistently matched according to the literature. To ensure accuracy, member checking through phone calls and emails with interviewees (Stake, 1995) was conducted after the interviews.

Throughout the stages, close attention was paid to patterns of consistency and contradictions. Such systematic analysis of data, as Plowman et al. (1995) argued, would provide a systematic process of analysis.

Findings and Discussion

The first research question examined how the CMP could enhance crisis planning.

The CMP Prompts Organizational Vigilance

As articulated by a respondent, "If what you are doing will kill people, then it is necessary to have a CMP" (personal communication, 2008). This highlights the paramount importance of the CMP as a powerful tool to facilitate the decision-making process in the face of severe and unexpected threats (McConnell & Drennan, 2006, p. 60). The importance of the CMP was further propagated by respondents who highlighted several key aspects of the plan that are important in a crisis. Despite its inherent limitations, the CMP remains a critical tool in mitigating a crisis.

Furthermore, it can be enhanced in several ways. First, the CMP provides a platform to identify key stakeholders. One respondent said, "In a crisis, the rule of thumb is to ensure that stakeholders are given pertinent information" (personal communication, 2008). This underlined stakeholder mapping was one of the primary uses of the CMP. Knowing the various stakeholders would enable the organization to tailor their messages appropriately. Another respondent argued that "organizations need to go through all the stakeholder groups to see which will be affected before deciding on their communication strategy" (personal communication, 2008). Organizations could also integrate feedback from stakeholders into their plans to ensure that their needs are adequately met. Apart from communicating information to the necessary stakeholder groups, the CMP also proved to be a useful document to help organizations to identify potential stakeholders who are victims of the crisis (personal communication, 2008). As highlighted by another respondent, "Crises breed secondary crises for stakeholders" (personal communication, 2008). This is especially so for multi-national corporations (MNCs) with many subsidiaries. Crises that occur to an organization in one country often have simultaneous impact with serious consequences in the other countries where the organization has a presence. Molleda, Connolly-Ahern, & Quinn (2005) termed this cross-national conflict shifting. Multi-national corporations with multiple partners or joint ventures are also not spared. The case of China's milk scandal aptly exemplifies this. The crisis of Sanlu, which was found to be producing contaminated milk and has since been folded, inevitably spilled over to Fonterra, a dairy organization from New Zealand that owns 43% of the former. As one of the stakeholders of San Lu, Fonterra has been unwittingly entrenched in the effects of this current scare (Ye, 2010).

Second, the CMP provides a platform to identify risk and issues. Issues are potential problems that may impact how the organization operates, while risks are vulnerabilities or weaknesses that organizations face (risks exist as part of business operations) that can be exploited or developed into crises (Coombs, 2007). As unpredictable as crises are, this emphasized the need to identify potential threats and identify issues before they actualize into full-blown crises. As one respondent said, doing so would "enable an organization to further plan for how to react in such instances" (personal communication, 2008). Furthermore, mapping out the possible risks ensures that disruptions to an organization's operations are minimized.

Third, the CMP provides a platform to delineate roles and responsibilities of the crisis management team. King (2002) proposed that for the team to better manage the crisis, it should (1) comprise members with prior interactions with one another; (2) be a heterogeneous group; (3) have members knowledgeable of the task at hand; and (4) possess a team leader who demonstrates a charismatic style of leadership. One respondent said that as the crisis unfolds, the "organization does not have the luxury of time to plan the rules and processes to solve a certain problem" (personal communication, 2008). Therefore, the team requires specific directions set beforehand to streamline the process of decision-making amidst the chaos (personal communication, 2008). These specific instructions are illustrated by the roles and responsibilities listed in the CMP, which help the team better frame their thought processes.

Fourth, the CMP provides the platform to ensure flow of communication during the crisis. Given that the organization is exhorted to constantly provide prompt and adequate information to their publics, the shifting of priorities during various stages of the crisis necessitates a constant stream of communication with different stakeholders at different times. One respondent said, "In the beginning, the organization's primary stakeholder might be the government. But in the later stages, the focus might shift and the organization would want to engage the people more" (personal communication, 2008). Therefore, organizations need to adopt different stances towards different stakeholders at different points in time, an idea which the Contingency Theory of Strategic Conflict Management posits (see Pang, Jin, & Cameron, 2010b). As explained by a respondent, knowing that one has to contact a stakeholder is basic crisis management. But knowing who to approach, how to approach the stakeholder, and at what point is another matter (personal communication, 2008).

Fifth, even as the CMP can be a useful but limited tool, developing a CMP is ultimately about developing a crisis preparedness mindset and resilience. As one respondent alluded, the exercise of developing "crisis response maps for several identified scenarios" (personal communication, 2008) is itself an exercise of developing a mindset. Similarly, the idea of diligently preparing documents is itself an exercise of developing resilience. Pauchant and Mitroff (1992) argued this may not always be the case in organizations. One way an organization shows the lack of such mindsets is to examine the way it rationalizes crisis efforts. An organization that regards planning as a time-consuming, profit-draining effort, for instance, will lack the temerity or fortitude to manage crisis when it hits.

Beyond examining the CMP, this study also sought to find out what situations would call for the activation of the CMP. While triggers can happen at different points and are subjected to a "case-by-case basis as no one issue or crisis is the same" (personal communication, 2008), two salient trigger points were identified amongst respondents. First, how seriously does the organization view the crisis? Referring to a recent case of tainted milk crisis in China in 2008, a respondent explained, "For organizations that take things very seriously, one death can be considered a crisis! Yet, there are these other organizations who disregard the magnitude of the issue and can simply remain unbothered by the real crisis and deaths going on around them" (personal communication, 2008).

Next, the activation of the CMP can also depend on crises faced by industry competitors. This is especially evident in the food or healthcare industry, whereby medical ailments faced by some consumers could signal all industry-players to raise their "red flags" (personal communication, 2008). One respondent cited the case in the product recall of the contact lens solution, Moisture Plus, in November 2006. The organization, Advanced Medical Optics (AMO), voluntarily issued a product recall during the "recall scare" period, whereby patients' eye infections were traced to products from AMO's competitor, Bausch & Lomb. This highlights the ability of a crisis to manifest into something of a larger magnitude involving other industry players. Fearn-Banks (2007) argued that a crisis has repercussions—not just on the organization but also on the industry as a whole. Importantly, organizations need to note that even if the crisis has not reached them, they have to be prepared to start taking action if the situation calls for it. Recognizing that the CMP serves an important function in an organization, respondents have also reiterated that having a CMP is not the only way to mitigate a crisis. As argued, organizations tend to endanger themselves by placing too much confidence on a CMP (Coombs, 2007). While having a plan helps an organization in the process of making decisions, organizations need to practice and conduct drills to constantly examine the relevance of the CMP (Pang, Cropp, & Cameron, 2006) in order to entrench a crisis preparedness mindset. This is reiterated by several respondents who claim that a CMP is only "as useful as the rehearsals you put into it" (personal communication, 2008), and the "most important thing in crisis preparation is to practice, practice and practice" (personal communication, 2008). Such training, coupled with regular updates to the manual, all culminate to having a plan that will prove helpful in a crisis.

As to when the CMP would be activated, respondents identified two trigger points: one, how serious the organization views the crisis and, second, what their competitors are doing. This supports what Fearn-Banks (2007) argued crisis to be: an event that affects not just specific organizations but the industry as a whole.

The second research question examined what other factors are involved in crisis planning besides the CMP.

Leadership in Establishing a Healthy Organizational Culture

The integrity of the management and empathy are some fundamental values respondents said should be embedded in the culture of any organization. Taking responsibility for a crisis was posited to be a characteristic that is extremely lacking in many organizations. One respondent said, "Management must learn that admitting fault and taking responsibility for the damage their organization has done is vital. Even though they may face the risk and danger of getting sued, they need to do the right thing" (personal communication, 2008). In explaining this point, the respondent highlighted the case of a national bank in Singapore that, in 2008, tried to shy away from blame when accused of not exercising enough care in the sale of financial products. This sensitive issue of admitting fault is still highly debatable though, as another respondent was also quick to state that "organizations should never admit fault until they are 120% sure that it is really their fault" (personal communication, 2008). Richard Branson, the founder of the Virgin Group, had a different perspective. He argued that even as the matter is under investigation, the CEO should take the lead in identifying where the internal processes had failed. He exhorted CEOs to apologize, express regret, describe what steps have been established to correct the mistake, and move on (Branson, 2010). To do so, the character of the leader, grounded on such core values as integrity, trust, truth, and human dignity, which impact on the leader's vision, ethics and behavior, is essential (Sankar, 2003). Jin (2010) argued that leaders ought to demonstrate emotional leadership by identifying constructive ways to connect with colleagues emotionally through understanding their "emotional and cognitive antecedents" (p. 160). Wooten and James (2008) argued that leaders need to demonstrate different competencies throughout the life cycle of a crisis, and one of which would be to create a conducive and inclusive culture.

Respondents argued that the culture of the organization, under the direction of the dominant coalition, could very well be the tipping point that makes or breaks a crisis. Corporate culture, the set of beliefs and values that are regarded and upheld as unifying forces that define an organization, are basic assumptions held by members about themselves and their environment (Pauchant & Mitroff, 1992). Culture can be operationalized as open or closed culture (Pang, Jin, & Cameron, 2010a) and may reflect or be a reflection of the values the dominant coalition holds. Though the corporate culture often reflects larger societal/business culture (Pauchant & Mitroff, 1992), it can also be defined by the leadership. Hagiwara (2007) found that the closed Japanese culture impeded Japanese organizations' ability to manage crises effectively. Pauchant & Mitroff (2006) further elaborated how some organizations are more crisis-prone as a result of their unhealthy cultures and found culture influenced organizational identity. They identified two types of organizations: self-inflated corporation (SIC) and positive self-regard corporation (PSRC). While their study concocted different measures on how these organizations would respond to challenges, like how they treat their employees for instance, as far as crisis is concerned, the former would regard crisis as caused by bad luck, while the latter would pre-empt crisis through strategic action.

This supports Marra's (2004) argument that practitioners cannot practice excellent crisis communications without a supportive organizational communication ideology. A communication culture, or ideology, is more likely to be shaped by management practices, or corporate culture. As a result of a conducive corporate communication culture, Marra (1998) found that the degree of communication autonomy practitioners enjoy to be a key factor in enhancing their effectiveness—not just in day-to-day situations but, more critically, during crisis situations. This would empower practitioners to assume leadership (Jin, 2010), and when they are empowered, they are likely to be able to carry out their functions more effectively and portray the organization in the best light they know how. This empowerment can be derived from endorsement and support by the dominant coalition and the corporate culture.

While the first stage of the Synthesis Model requires ongoing PR efforts, the culture of an organization is the larger, overarching framework that sets the direction an organization will take in mitigating crises. Displaying values such as integrity and empathy in dealing with issues, culture creates the *social glue* that bonds the internal stakeholders and instils a responsible image within the organization. More than that, this also creates a cohesion, which will ultimately produce employees who are, in fact, "the best spokespersons for their organizations" (personal communication, 2008). Thus, when a crisis occurs, the entire organization can then function as a single entity and portray a united image that will serve to bolster the organization's reputation, as was seen in JetBlue. This overarching characteristic of the organization should thus be in place, even before a crisis occurs.

Proactive Communication with Stakeholders

As argued, the CMP provides a platform for stakeholder mapping. An empowering organizational culture encourages employees to embark on proactive communication with stakeholders. As Marra (2004) argued, when a crisis strikes, vital information needs to be disseminated to all who are affected and potentially affected by it. Several respondents highlighted the need to "communicate fast and communicate often" (personal communication, 2008) to acknowledge the situation—even if there is no information at hand. One respondent shared his experience managing the Severe Acute Respiratory Syndrome (SARS) crisis that gripped part of Asia and North America in 2003. The respondent was working in Singapore at the time and explained that due to the fact that physical contact was kept to a minimum to avoid spreading of the disease, communication was vital to keep hospitals and medical staff constantly updated on the progress of the efforts undertaken to contain the virus. To facilitate this process, newsletters were delivered twice a day to give the latest updates and information on new deaths and new strains found (personal communication, 2008). One reason for this proactive communication was the enabling culture practiced and led by the regulatory authorities at that time: *Let's be open that this disease has reached our shores; let's be proactive in dealing with it; and one way to deal with it is to communicate proactively with our stakeholders.*

Another respondent recounted his experience helping a food manufacturer manage a contamination crisis. When investigations identified the contamination of cakes within its factory causing more than 200 cases of food poisoning, this led to a series of product recalls, along with the shutting down of the organization's manufacturing facility and 38 franchises. Yet, the organization managed to turn the crisis around through proactive communication. "They opened their books to the public to maintain transparency, and constantly made announcements to prove that they meant what they said. They even invited the media down to view and write stories on how they cleaned their outlets, and kept the media updated on the changes made" (personal communication, 2008). The organization also set up a hotline to address the concerns of affected publics and worked very closely with the regulatory authority to ensure the quality standards of the products in their shops met benchmarks.

As a result, the chain's overall sales figures increased by up to 15%, signalling that business was not only back to normal but even better than before. Ironically, the cakes that were contaminated became best sellers again in some outlets. With its media-centric strategy, the food manufacturer successfully shifted public opinion in their favor. Proactive communication, however, should not just take place during a crisis. The bigger lesson learned is that proactive communication should take place before crisis hits (Fearn-Banks, 2007) so that should crisis hit, the foundation of speaking and relating with stakeholders is already in place to re-establish relations.

During a crisis, stakeholders all want to know what is going on and how the crisis could affect them. Thus, organizations should not only delve into the mechanics of resolving the crisis at hand; they should also seek to reach out to their affected stakeholders and engage them. Even if they have no information at hand, maintaining open communication with their stakeholders is imperative, as a lack of information could lead to a loss of confidence in the organization. As stated by a respondent, "A crisis is a crisis of confidence" (personal communication, 2008). Ultimately, proactive external communications is also needed to help the public regain their initial confidence in the organization and to dispel any notions of evasiveness that the public has of the organization.

Respondents said proactive communication could take place in different forms. Besides direct communication, the most visible would be through the media. While traditional media are important, the use of new media, as demonstrated by American Airline JetBlue's apology through the organization's website and on a YouTube video, exemplified an alternative avenue that organizations can explore in mitigating a crisis. One respondent said that the online video, which showcased the CEO apologizing to all affected parties for an operational meltdown, which began with an ice storm that subsequently caused the cancellation of more than 1,000 flights, even as hundreds of passengers were trapped for hours in idled planes ("CEO of

JetBlue is truly sorry—clearly he is," 2007), showed JetBlue's sincerity to reach out to their stakeholders and, more importantly, allowed it to be viewed at a time when the entire organization was extremely busy in the resolving of this crisis and could hardly be reached (personal communication, 2008).

Towards a Revised Synthesis Model

Building on the points highlighted above, more factors were identified to be crucial in ameliorating a crisis. While the CMP provides detailed instructions and directions for an organization to take when a crisis strikes, there should be concurrent efforts among all stakeholders that will ultimately culminate in an integrated plan for organizations to rely on.

This chapter therefore seeks to expand the current limitations of Horsley and Barker's (2002) Synthesis Model of Crisis Communication by positing three new stages: (1) propagating a healthy organizational culture, (2) encouraging proactive external communications, and (3) activating the CMP.

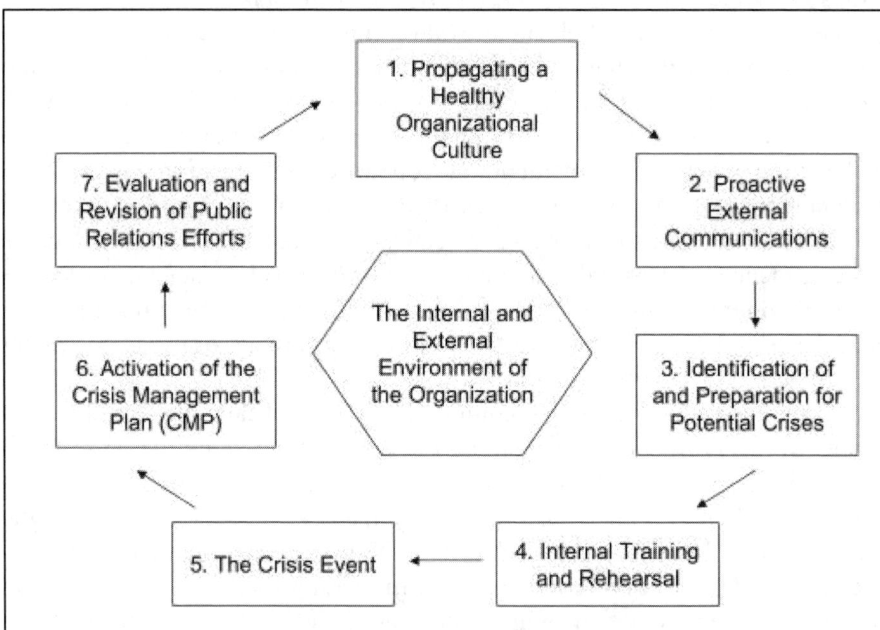

Figure 2: **Revised Synthesis Model of Crisis Communication**

While we recognize ongoing public relations efforts are important tasks to be recommended as stage one in Horsley and Barker's (2002) model, we argue that prior to this engagement two factors will determine the effectiveness of such efforts. These two, we argue, would be a healthy organizational culture and, in turn, a culture of engaging in proactive communication. This would provide the overarching strategy in ongoing public relations efforts and ensuring that continuous attention is paid to stakeholders and that communication systems are in place to help the organization get its messages out with a view of building strong relations with stakeholders before any crisis hits. This would set the organization on firm footing as it embarks on the identification of and preparation for potential crises through the conceptualization of the CMP. Besides detailing the procedure, process, and people that would be activated when crisis hits, our interviews found that the CMP would provide the platform for organizations to identify key stakeholders; identify risks and issues; delineate roles and responsibilities of the crisis management team; and develop a system to ensure communication flow—all with a view of building crisis resilience. This is followed by Horsley and Barker's (2002) stage on internal training and rehearsal. When the crisis event occurs, organizations should be instructed on when to activate the CMP. This is followed by Horsley and Barker's (2002) evaluation and revision of public relations efforts.

Conclusion

This study has examined how the CMP could enhance crisis planning and what other factors are involved in crisis planning using the Synthesis Model of Crisis Communication (Horsley & Barker, 2002) as its theoretical lens. It is an exploratory study to identify more factors involved in crisis preparedness that seeks to understand where the CMP fits into overall planning. Since this is arguably one of the first studies building on the Synthesis Model, it sheds new insights, which may help to take the model forward. The findings from this study provide the initial framework for application and testing to make it more rigorous in both public and private sectors.

One limitation of this study is interviews with public relations consultants rather than in-house practitioners. This is due in part to the sensitivity of the topic, which limits the amount of information that the study can garner from in-house practitioners, whom Pang, Cropp and Cameron (2006) argued may be hesitant to grant access because of the highly sensitive nature of the topic. To overcome this, we chose consultants who had in-house experience and who are also consulting with organizations who can accord us views from both perspectives. Admittedly, the small interview pool may also limit the range of perspectives. However, given the exploratory nature of this study, it is a limitation we accept.

Most importantly, we believe this will add further understanding of the components that comprise a rigorous model of crisis communication. We hope this conversation will continue where we have left off.

References

Aronson, J. (1994). A pragmatic view of thematic analysis. *The Qualitative Report, 2*(1). Retrieved from http://www.nova.edu/ssss/QR/BackIssues/QR2-1/aronson.html

Barton, L. (1994). Crisis management: Preparing for and managing disasters. *Cornell Hotel and Restaurant Administration Quarterly, 35*(2), 59–66.

Borda, J. L., & Mackey-Kallis, S. (2004). A model for crisis management. In Dan P. Millar & Robert L. Heath (Eds.), *Responding to crisis: A rhetorical approach to crisis communication* (pp. 117–137). Mahwah, NJ: Lawrence Erlbaum.

Boyatzis, R. E. (1998). *Transforming qualitative information: Thematic analysis and code development.* London, UK: Sage.

Branson, R. (2010). What a CEO should do when disaster strikes, *Today*, p. 17.

Burnett, J. J. (1998). A strategic approach to managing crisis. *Public Relations Review, 24*(4), 475–485.

Campbell, R. J., & Sato, H. (2009). Examination of a global prohibition regime: A comparative study of Japanese and US newspapers on the issue of tobacco regulation. *The International Communication Gazette, 71*(3), 161–179.

CEO of JetBlue is truly sorry—clearly he is. (2007, Feb. 24). *Oakland Tribune.* Retrieved October 24, 2008, from http://findarticles.com/p/articles/mi_qn4176/is_20070224/ai_n18638813

Chay-Nemeth, C. (2003). Becoming professionals: A portrait of public relations in Singapore. In K. Sriramesh & D. Verčič (Eds.), *The global public relations handbook* (pp. 86–105). Mahwah, NJ: Lawrence Erlbaum.

Cloudman, R., & Hallahan, K. (2006). Crisis communications preparedness among U.S. organizations: Activities and assessments by public relations practitioners. *Public Relations Review, 32*, 367–376.

Coombs, W. T. (2000). Designing post-crisis messages: Lessons for crisis response strategies. *Review of Business, 21*(3), 37–49.

———. (2007). *Ongoing crisis communication: Planning, managing, and responding.* Thousand Oaks, CA: Sage.

Fearn-Banks, K. (2007). *Crisis communications* (3rd ed.). Mahwah, NJ: Lawrence Erlbaum.

Fink, S. (1986). *Crisis management: Planning for the inevitable.* New York: AMACOM.

Freeman, R. E. (1984). *Strategic management: A stakeholder approach.* Marshfield, MA: Pitman Publishing.

Freeman, R. E., & Gilbert, D. R. (1987). Managing stakeholder interests. In S. P. Sethi & C.M. Fable (Eds.), *Business and society: Dimensions of conflict and cooperation* (pp. 379–422). Lexington, MA: Lexington Books.

Gonzalez-Herrero, A., & Pratt, C. B. (1996). An integrated symmetrical model for crisis communication management. *Journal of Public Relations Research, 8*(2), 79–105.

Hagiwara, T. (2007). The eight characteristics of Japanese crisis-prone organizations. In C. M. Pearson, C. Roux-Dufort, & J. A. Claire (Eds.), *International handbook of organizational crisis management* (pp. 252–270). Thousand Oaks, CA: Sage.

Heath, R. L. (1997). *Strategic issues management: Organizations and public policy challenges*. Thousand Oaks, CA: Sage.

Horsley, J. S., & Barker, R. T. (2002). Towards a Synthesis Model for Crisis Communication in the public sector: An initial investigation. *Journal of Business and Technical Communication, 16*(4), 406–440.

Jin, Y. (2010). Emotional leadership as a key dimension of public relations leadership: A national survey of public relations leaders. *Journal of Public Relations Research, 22*(2), 159–181.

Jin, Y., & Pang, A. (2010). Future directions of crisis communication research: Emotions in crisis—The next frontier. In W. T. Coombs & S. J. Holladay (Eds.), *Handbook of crisis communication* (pp. 677–682). Malden, MA: Wiley-Blackwell.

Jin, Y., Pang, A., & Cameron, G. T. (2007). Integrated crisis mapping: Towards a publics-based, emotion-driven conceptualization in crisis communication. *Sphera Publica, 7,* 81–96.

———. (2009). *Toward a publics-driven, emotion-based system in crisis communication: Unearthing dominant emotions in multi-staged testing of the Integrated Crisis Mapping (ICM) Model.* Boston, MA: AEJMC.

King, G. (2002). Crisis management and team effectiveness: A closer examination. *Journal of Business Ethics, 41*(3), 235–250.

Knapp, M. L., Miller, J. L., & Fudge, K. (1994). *Handbook of interpersonal communication.* London: Sage.

Lerbinger, O. (1997). *The crisis manager.* Mahwah, NJ: Lawrence Erlbaum.

Lim, S., Goh, J., & Sriramesh, K. (2005). Applicability of the general principles of excellent public relations in a different cultural context: The case study of Singapore. *Journal of Public Relations Research, 17*(4), 315–340.

Liu, B. F., & Horsley, J. S. (2007). The government communication decision wheel: Toward a public relations model for the public sector. *Journal of Public Relations Research, 19*(4), 377–393.

Marra, F. J. (1998). Crisis communication plans: Poor predictors of excellent crisis public relations. *Public Relations Review, 24*(4), 461–474.

———. (2004). Excellent crisis communication: Beyond crisis plans. In Millar, D. P. & Heath R. L. (Eds.), *Responding to crisis: A rhetorical approach to crisis communication* (pp. 311–325). Mahwah, NJ: Lawrence Erlbaum.

McConnell, A., & Drennan, L. (2006). Mission impossible? Planning and preparing for crisis. *Journal of Contingencies & Crisis Management, 14*(2), 59–70.

Mitroff, I. I., & Anagnos, G. (2001). *Managing crises before they happen.* New York: AMACOM.

Molleda, J. C., Connolly-Ahern, C., & Quinn, C. (2005). Cross national conflict shifting: Expanding a theory of global public relations management through quantitative content analysis. *Journalism Studies, 6*(1), 87–102.

Olaniran, B. A., & Williams, D. E. (2001). Anticipatory model of crisis management. In R. L. Heath (Ed.), *Handbook of public relations* (pp. 487–500). Thousand Oaks, CA: Sage.

Pang, A., Cropp, F., & Cameron, G. T. (2006). Corporate crisis planning: Tensions, issues, and contradictions. *Journal of Communication Management, 10*(4), 371–389.

Pang, A., Jin, Y., & Cameron, G. T. (2010a). *Organizational strategies to engage publics' emotions in multistaged testing of the Integrated Crisis Mapping Model.* Singapore: ICA.

———. (2010b). Contingency theory of strategic conflict management: Directions for the practice of crisis communication from a decade of theory development, discovery and dialogue. In W. T. Coombs & S. J. Holladay (Eds.), *Handbook of crisis communication* (pp. 527–549). Malden, MA: Wiley-Blackwell.

Pang, A., & Yeo, S. L. (2010). *Crisis management consultancy: Exploration of the expertise, experience and expediency of consultants in public relations agencies in Singapore.* Hong Kong: CCA.

Pauchant, T. C., & Mitroff, I. I. (1992). *Transforming the crisis-prone organization.* San Francisco: Jossey-Bass.

———. (2006). Crisis-prone versus crisis-avoiding organizations: Is your company's culture its own worst enemy in creating crises? In D. Smith & D. Elliott (Eds.), *Key readings in crisis management* (pp. 136–146). NY: Routledge.

Pearson, C. M., & Mitroff, I. I. (1993). From crisis prone to crisis prepared: A framework for crisis management. *Academy of Management Executive*, 7(1), 48–60.

Penrose, J. M. (2000). The role of perception in crisis planning. *Public Relations Review, 26(2),* 155–171.

Pinsdorf, M. K. (1987). *Communicating when your company is under siege: Surviving public crises.* Lexington, MA: D. C. Heath and Company.

———. (1995). Flying different skies: How cultures respond to airline disasters. *Public Relations Review, 17*(1), 37–56.

Plowman, K. D., Re Velle, C., Meirovich, S., Pien, M., Stemple, R., Sheng, V., & Fay, K. (1995). Walgreens: A case study in health care issues and conflict resolution. *Journal of Public Relations Research, 7*(4), 231–258.

Sankar, Y. (2003). Character not charisma is the critical measure of leadership excellence. *Journal of Leadership & Organizational Studies, 9*(4), 45–55.

Seeger, M. W., Sellnow, T. L., & Ulmer, R. R. (1998). Communication, organization and crisis. *Communication Yearbook, 21,* 231–275.

———. (2003). *Communication and organizational crisis.* Westport, CT: Praeger.

Stake, R. E. (1995). *The art of case study research.* Thousand Oaks, CA: Sage.

Tiller, M. J. (1994, April). Is your disaster plan effective? *Management Review*, p. 57.

Ulmer, R. R. (2001). Effective crisis management through established stakeholder relationships: Malden Mills as a case study. *Management Communication Quarterly,* 14(4), 590–615.

Weick, K. E. (2001). *Sensemaking in organizations.* Blackwell: Oxford.

Wooten, L. P., & James, E. H. (2008). Linking crisis management and leadership competences: The role of human resource development. *Advances in Developing Human Resources, 10*(3), 352–379.

Ye, L. (2010). *Understanding the Chinese way of crisis management: A case study of crisis communication of Sanlu Group in 2008.* Master's dissertation, Wee Kim Wee School of Communication and Information, Nanyang Technological University. Unpublished manuscript.

Section 3

Managing Diversity in Crisis

Chapter 10

Lost in Translation: How Emergency Managers Communicate Preparedness with Vulnerable Populations

Shari R. Veil, Clara Mitcham & Timothy L. Sellnow

This chapter is a must-read for all who are interested in emergency management procedures. The authors begin by highlighting the communication failures during Hurricane Katrina. The lack of understanding and preparation witnessed in 2005 help establish the need for the study seen in this chapter. Thirty-seven emergency managers were interviewed concerning their preparations for responding to communities with diverse populations. Interviewees addressed topics of communication tactics they used, supporting agencies, communication with and through community groups and leaders, and the use of translators. The experts also addressed a range of barriers they encounter when communicating with members of diverse populations.

Introduction

Hurricane Katrina pulled back the curtain on emergency management in the United States. While much attention was drawn to the delayed and seemingly incompetent response system, bureaucratic infighting, and blame shifting, Hurricane Katrina also vividly exposed our nation's vulnerability to disasters. Even more, Katrina exposed our failures in accounting for the most vulnerable citizens in any crisis. The elderly, children, and individuals with disabilities and medical or other special needs are often identified as vulnerable populations (Uscher-Pines et al., 2009). But the vulnerable populations list was expanded when Hurricane Katrina devastated areas largely populated by low-income and impoverished individuals living in substandard housing, particularly African Americans (Andrulis, Siddiqui, &

Gantner, 2007). This population faced additional obstacles in following the evacuation order from New Orleans including limited access to transportation, a distrust of government figures, and inconsistent access to and viewing of traditional media (Lachlan, Burke, Spence, & Griffin, 2009). Latinos and Asian Americans faced similar barriers (Mufliz, 2006; NCAPA, 2005), "compounded by issues of language, culture, and their status as undocumented or uninsured residents" (Andrulis et al., 2007, p. 1270). Issuing a mandatory evacuation order in a community with a substantial portion of the population lacking money or transportation to evacuate—communicated by an untrusted government spokesperson, through media channels to which many residents had little or no access, and in a language not understood—clearly demonstrated the cultural deafness of the U.S. emergency management system in 2005.

In the half decade since Hurricane Katrina, scholars and government agencies have focused a great deal of attention on these and other barriers that prevent populations from receiving risk communication messages in time to limit the disproportionate effects of crises (Messias & Lacy, 2007). "The circumstances surrounding Hurricane Katrina taught communication practitioners several important lessons, arguably the most important of which is the need for effective risk communication messages in the period before a natural disaster" (Lachlan et al., 2009, p. 296). Crafting appropriate messages is less of a quandary than getting individuals to take the protective actions emphasized in the messages. Encouraging people to prepare for uncertain and upsetting events such as emergencies and disasters is particularly challenging (Littlefield et al., 2010; Veil, Littlefield, & Rowan, 2009).

The pre-crisis communication of the Federal Emergency Management Agency (FEMA) and the overarching Department of Homeland Security (DHS) has focused on encouraging individuals, communities, and businesses to develop personal emergency plans and stock emergency kits. All U.S. residents are recommended to stock household emergency kits that include water and food for three days, a battery-powered radio, flashlight, first aid kit, and other necessities such as prescription medicines, baby supplies, local maps, and can openers (FEMA, www.ready.gov). These kits are designed to prepare residents to sustain themselves in a naturally occurring disaster or intentional terrorist attack until emergency help can arrive. FEMA Administrator Craig Fugate stated that "personal disaster preparedness is and must be a national priority," and in fact, "nothing will contribute more to saving and sustaining lives than a citizenry prepared and provisioned to live in a reduced-services environment in the days immediately following a catastrophic disaster" (FEMA, 2009, p. 9).

As part of their many responsibilities, local emergency managers have been directed by the federal government to increase community members' preparedness for emergencies by encouraging community residents to develop emergency plans and stock emergency kits. This study examines the efforts of local emergency managers to communicate preparedness messages with vulnerable populations. Specifically, 37 emergency managers from eight states were interviewed to discern the current actions taken to increase emergency kit preparedness with diverse community members. First, the intermix of risk and crisis communication in communicating pre-crisis messages with vulnerable populations is explained. Second, barriers to reaching diverse populations with risk messages are described. Third, themes related to communicating preparedness with vulnerable populations are identified and described. Finally, conclusions are drawn to assess emergency managers' current communication strategies for and barriers to increasing pre-crisis preparedness of vulnerable populations.

Pre-crisis Communication

Pre-crisis communication is a combination of risk and crisis communication. Risk, in this context, is contingent upon a crisis occurring. Without a crisis, there is no manifested risk. Risk communication is defined as "the intentional effort to inform the public about risks and persuade individuals to modify their behavior to reduce risk" (Seeger & Reynolds, 2007, p. 9). Thus, encouraging individuals to prepare emergency kits and plans to limit risks they may face in a crisis fits well within the realm of risk communication. Risk and crisis communication are highly interrelated such that risk communication before a crisis influences perceptions, expectations, and behavioral responses to communication during a crisis. Both risk and crisis communication also "share an essential purpose of seeking to limit, contain, mitigate, and reduce harm" (Reynolds & Seeger, 2005, p. 48). However, "the persuasive nature of risk communication and time constraints of crises have fostered distinct research traditions" (Veil, Reynolds, Sellnow, & Seeger, 2008, p. 27)

Risk Communication

Much of the research on how to reach diverse publics with risk messages has been centralized in health communication campaign design. In the health context, personal risks exist regardless of a pending crisis. In hopes of persuading individuals to make personal behavioral changes that reduce health risks, campaign research has focused on risk communication messages designed specifically to match audience needs, values, background, culture, diversity, health literacy, and experience (Murray-Johnson et al., 2001; Institute of Medicine, 2002). Health communication research revealed that risk levels are uneven across diverse populations, and unfortunately, often those with the highest frequency of health burdens have the least access to the essential "information, communication technologies, health care, and supporting social services" (USDHHS, 2000, p. 9).

Crisis Communication

A traditional assumption of crisis communication practice and research is that because of the immediate need for information in a crisis, there typically is not time to tailor messages to diverse publics. "The immediate communication needs are to reduce the uncertainty, allowing audiences to create a basic understanding of what happened so that they may act appropriately" (Reynolds & Seeger, 2005, p. 50). Crisis communication includes providing information for what to do to protect oneself physically, adjusting information to assist individuals in coping psychologically with the crisis, and reputation management responses that seek to protect the reputation of the organization or agency involved (Coombs, 2007). Indeed, much of the research on crisis communication has been on restoring organizational reputation following crisis.

Emergency Communication

The Centers for Disease Control and Prevention (CDC) endeavored to combine risk and crisis communication concepts in an integrated framework called Crisis and Emergency Risk Communication (CERC). The CERC model focuses on the audiences' informational needs during an emergency (Reynolds & Seeger, 2005). This orientation places an obligation on the communicators, in this case emergency managers, to understand the complex needs, background, and culture of the intended audiences before the crisis occurs. This approach represents a shift from the discredited linear approach where authorities were thought to have control over the communication process and could send messages directly to audiences through mass media channels, and the audiences would absorb the messages (Botan & Penchalapadu, 2008). Previous research on emergency manager communication practices demonstrated a similar, one-way approach by which emergency managers thought dissemination of the message meant the audiences should automatically act in the recommended fashion (Veil et al., 2009).

In striving for a more participatory communication model, one of the key themes of CERC is understanding audience diversity. "Given that audience diversity during a crisis often translates into uneven vulnerabilities, understanding these variables within the crisis context is a particularly important area of research" (Veil et al., 2008, p. 31). After all, the publics' response to emergencies can "only be as good as the information that people are given" (Alexander, 2002, p. 148). But, as noted, just because the information is disseminated, that does not mean the public is receiving or understanding the message. The following section describes barriers to pre-crisis risk messages, especially in vulnerable populations.

Communicating with Vulnerable Populations

Vulnerable populations were previously defined by FEMA as individuals with physical, medical, sensory, or other cognitive disabilities (FEMA, 2004). Additional vulnerable populations now identified include the elderly, minorities, non-English speaking individuals, people without vehicles, and people with special dietary needs. People of lower socioeconomic status are also a growing vulnerable population within the United States. In the case of disasters, low-income households are at a disadvantage because there is generally less insurance coverage, less physically stable housing, and fewer social services due to low tax bases (Zakour & Harrell, 2003). These same issues also stunt the recovery process, leaving communities and individuals financially and psychologically devastated (Zakour & Harrell, 2003). People living in poverty are generally unable to afford supplies recommended for an at-home kit and may be unable to gain access to emergency preparedness information because they do not have access to information technology. Also, they often are not self-sufficient in transportation.

Racial and Ethnic Diversity

Racially and ethnically diverse communities are vulnerable to disaster—not because of skin color, but because, statistically, they are disproportionally represented among individuals with a lower socioeconomic status. They also have different perceptions of risk and use different communication channels (Lachlan et al., 2009). In the wake of the anthrax attacks following 9/11, studies revealed distrust of government messages by racial minority groups (Quinn, Thomas, & McAllister, 2005). Minority stakeholders seek out information in different ways (Spence, Lachlan, & Griffin, 2007), and generalized messages are less effective in motivating specific subpopulations (Andrulis et al., 2007). Andrulis et al. (2007) surmise that the factors contributing to the vulnerability of racial/ethnic minorities include "socioeconomic differences, culture and language barriers, lower perceived personal risk from emergencies, distrust of warning messengers, lack of preparation and protective action, and reliance on informal sources of information" (p. 1270). Previous research suggests that racial and ethnic minorities are less likely to accept a risk or warning message from a government representative as credible without confirmation of the message from within their interpersonal networks (Lindell & Perry, 2004). Thus, during Hurricane Katrina there may have existed fewer credible outlets to influence African Americans about the severity of the threat (Lachlan et al., 2009).

Sellnow and Vidoloff (2009) propose that acknowledging and accounting for cultural differences should be considered a best practice of risk and crisis communication. They suggest that crisis communicators should make additional efforts to reach underrepresented populations including those who are enduring poverty, are new to America, or others who have limited access to mainstream media. Understanding how messages may be interpreted in different cultures, ethnic groups, and socioeconomic groups is essential, as is determining the mass communication and interpersonal communication channels best suited to reach them. Blanchard et al. (2005) contend that vulnerable populations should be specifically targeted in an emergency and require tailored approaches to ensure effective communication.

One of the criticisms leveled at the Red Cross after Hurricane Katrina focused on the organization's perceived lack of cultural awareness. In his testimony before a subcommittee of the United States House of Representatives, Kevin Brown, the chief operating officer of Red Cross, summed up these criticisms:

> During the relief effort, some advocacy groups observed that the Red Cross had not consistently met the needs of a diverse segment of their communities [...] that the Red Cross workforce lacked diversity from top to bottom, and, as a result, was not sufficiently sensitive to racial and cultural issues [...] that the Red Cross lacked "cultural competence" in its response to Katrina. Other groups reported Red Cross communication failures with minority populations, particularly in remote areas along the Gulf Coast. Another recurring complaint was that the Red Cross had not done enough before Katrina struck to foster the necessary relationships and develop agreements with local and national organizations that serve diverse populations. (Brown, 2007, para. 14)

In short, the Red Cross failed to account for cultural differences in its crisis response (Veil & Husted, in press). That said, however, the organization has since taken steps to improve its cultural competence, at least by forming partnerships with other organizations that work with diverse publics (Brown, 2007).

While most noticeable in the aftermath of Hurricane Katrina, African Americans were not the only minority deeply impacted by the crisis. A study of the effect of Hurricane Katrina on the Vietnamese population of New Orleans, which also suffers from a language barrier, showed that low-income and ethnic minority status directly correlate with higher financial stress, greater exposure to environmental elements, and high levels of post-traumatic stress syndrome (Chen et al., 2007). Chen et al. (2007) suggest, however, that the role of social support groups, especially churches, helped in the post-crisis communication, as many church and religious organizational leaders served as translators between government entities and the Vietnamese community.

Language Diversity

Communication with non-English speaking ethnic minorities is often compounded by the fact that a notable portion of the minority population who do not speak English are also of lower socioeconomic status. Therefore, not only is this group unable to understand messages about crisis preparedness, but they also lack the resources needed for responding to the degree of the average American citizen. With one-fifth of the population now speaking a language other than English, the number of non-English speaking and non-fluent English speaking ethnic minorities in the United States has increased significantly in the past quarter century (Shin & Bruno, 2003). While much of this increase may be attributed to a rise in the number of Spanish-speaking immigrants, more non-English speaking groups from all backgrounds, falling under several different categories of citizenship, are establishing permanent or semi-permanent residency in the United States (Shin & Bruno, 2003). The 2000 Census indicates that there were almost 47 million people living in the United States whose first language is not English and who have varying levels of English comprehension. Of the 47 million, over 7.5 million said they did not speak English very well, and 3.5 million did not speak English at all (Shin & Bruno, 2003).

Current disaster and emergency response planning does not adequately address the needs of limited English proficient communities (Shiu-Thornton et al., 2007). FEMA, DHS, and the Red Cross do have literal translations of preparedness messages available, mostly in Spanish. However, few current initiatives address the needs of racial/ethnic groups requiring attention to culture, literacy, and trust (Andrulis et al., 2007). Looking only at the needs of Spanish-speaking individuals, Fisher Liu (2007) found that 70% of the 435 emergency manager respondents surveyed do employ at least one channel to communicate Spanish-language disaster information; however, on average, they employ far fewer channels to communicate Spanish-language disaster information (three) than English-language disaster information (nine). A case study of Hurricane Andrew revealed that when seeking financial help after the hurricane, many non-English speakers reported that they did not seek aid from FEMA because they did not understand the process and could not find any Spanish-speaking FEMA representatives (Cherry & Cherry, 1997).

Assumptions of Preparedness Messages

Designing messages to target diverse populations requires detailed local knowledge of those at risk (Parker & Handmer, 1998). "Emergency managers often assume that all people are able to receive and understand their disaster warnings, plans, and information" (Peguero, 2006, p. 6), but the dissemination of messages through traditional media channels does not take into account racially and ethnically diverse communities. Most resources and materials targeting minorities are disseminated primarily through the Internet (Andrulis et al., 2007). "Unfortunately, many racial/ethnic groups might not benefit from these resources because of limited ability to access the Internet and limited skills to navigate complex Web-based systems predominantly in English" (p. 1272). Andrulis et al. (2007) suggest the primary channels for communicating with racial and ethnic groups should include "media controlled by racial/ethnic groups, faith-based organizations, community leaders, and community-based outreach workers" (p. 1275).

Many current preparedness guidelines and policies are based on the expectation that each U.S. citizen has a civic responsibility to maintain a certain level of preparedness (Veil, 2008); however, as the population becomes increasingly more diversified, there are a growing number of vulnerable populations that either cannot understand the preparedness messages available or do not receive essential information about emergency preparedness (Andrulis et al., 2007). As shown in this literature review, researchers are developing communication models to include elements of risk and crisis communication and account for diverse audience needs. But in practice, limited progress has been made in preparing vulnerable populations for crisis.

Research indicates that many people in the United States do not yet have emergency kits or plans, despite efforts to encourage these behaviors (CEG, 2006; MACRO, 2006). For example, one study found that only 36% of respondents had household emergency kits, and only 27% had a designated place where family members could meet if a disaster occurred (CEG, 2006). Individuals with disabilities and medical and other special needs are not more likely to prepare a complete emergency kit and plan (Uscher-Pines et al., 2009), and research was not located depicting the preparedness level of other vulnerable populations. This study looks specifically at how emergency managers communicate pre-crisis information with vulnerable populations. In total, 37 emergency managers from eight states were interviewed to determine current strategies for and barriers to communicating preparedness with vulnerable populations. The following methods describe the process of data collection and analysis.

Method

Participants of this study were local government or public university employees with responsibility for local emergency preparedness. The job titles of those interviewed varied (e.g., emergency management director, assistant emergency manager, public affairs director, public health director, etc.). Each individual interviewed was eligible to be a member of his or her local emergency planning committee (LEPC), and most worked for a county or city. Selected individuals who were interviewed worked in towns, cities, or counties near the project authors and a team of interviewers coordinated by the National Communication Association.

Participants were identified through a purposive sampling technique. They were located through local government websites identifying those responsible for emergency management. These individuals were contacted through an e-mailed or mailed letter requesting their participation in an interview. All interviewers received approval from their respective university Institutional Review Boards. When individuals agreed to participate, they were contacted at a specified time, date, and location. Interviews lasted for about an hour and were conducted face-to-face in locations selected by the participants, typically at participants' offices.

Answers to interview questions were transcribed by interviewers who took extensive notes and typed transcripts immediately after interview completion. Participants were identified only by a code number (e.g., OK

EM 1), so investigators were aware of the state but not the identity of the interviewees. Because some natural disasters, accidents, and terrorist attacks have already occurred or are more common to certain regions of the United States, it may be possible to infer the regions of the country being discussed by participants. Transcripts were analyzed to identify patterns that could guide research on enhancing local preparedness for emergencies. In this study, we describe responses to a subset of the questions related to efforts taken to communicate preparedness messages to vulnerable populations in the emergency managers' jurisdictions.

Results

Emergency managers were interviewed to discern the current actions taken to increase emergency kit preparedness with diverse community members. While most emergency managers agree that communication with vulnerable populations is important, few are taking additional steps to ensure preparedness messages are received and acted upon. In general, the emergency managers felt they needed more resources to communicate during a crisis, while preparedness communication was not as high of a priority. Many emergency managers said vulnerable populations did not constitute a large enough portion of the community, and/or effective communication would be impossible due to lack of resources. Research is not being conducted by emergency managers to determine if vulnerable populations have emergency kits. In fact, evaluation of communication strategies for any publics was not reported by any of the emergency managers.

Emergency managers described what communication strategies they are currently using to reach vulnerable populations as well as barriers to communicating with vulnerable populations. Subthemes were identified under the broad categories of communication strategies and communication barriers.

Communication Strategies

Subthemes of communication strategies identified in the transcripts included efforts to reach vulnerable populations through communication tactics, reliance on supporting agencies, involvement of community groups and leaders, and having translators available in a crisis response.

Communication Tactics

Most commonly, emergency managers described tactics for reaching non-English speaking communities. Preparedness information is printed or made available online in other languages, primarily Spanish. Approximately one-third of the emergency managers explicitly stated that their organization has bilingual or multi-lingual printed material available. Of the counties with this type of material available, three managers said they had information in languages other than Spanish. "As we develop materials, especially written materials, something that I hope we can do is minimally put things in Spanish. We have a very large Hispanic community. I don't know if we can economically…communicate in every language we have in the city" (VA EM 6). The effective translation of a number of these materials is also a concern of emergency managers. "The city, through Public Health has worked with translators to translate the preparedness guide into a number of different languages. Public Health also tries to get info to leaders in the ethnic communities, and the directions for forms are now in different languages" (ND EM 5). Alternatives to online and printed resources include radio and television public service announcements. "In the past, I have done presentations on the city/county Spanish/Latino radio station in the city. We have done presentations on some Christian radio stations" (PA EM 2).

Reliance on Supporting Agencies

Due to the lack of resources available to some emergency managers, there is a strong reliance on other national, state, and local organizations to provide and even distribute preparedness information. For example, three emergency managers said that they rely on the bilingual preparedness guides available through the Red Cross to reach target populations: "…free Red Cross brochures have been available in Spanish and disbursed in various public locations (i.e., library, county fair)" (OK EM 5). Also, more diverse communities with large populations that speak languages other than Spanish rely on the Red Cross for translated information. "We are [translating information] for larger ethnic groups, but we have a large Vietnamese and Asian community and we don't have materials. We can access the Red Cross version of our stuff and they have all the languages" (TX EM 2).

Beyond the Red Cross, emergency managers use and rely on other organizations to distribute information about preparedness because they cannot produce or are unable to translate messages. One city with large immigrant and refugee populations relies on resettlement agencies and other social service providers to pass along, and in some cases translate, printed materials from the local government (PA EM 1). Another manager sees these agencies as a tool that will save them time and resources. "For any ethnic minority, there are agencies or resources that are already in place, maybe the healthcare community, social services community that speaks the language" (ND EM 4). Some counties have partnerships in place to expedite communication. By working with local Lutheran Social Services, the Learning and Giving Center, and the State Refugee Advisory Council, one county has been able to form a communication network (ND EM 1).

Community Groups and Leaders

Other local resources are available in the form of churches, community groups, and individual community leaders. Emergency managers indicated that the use of these groups for communication about preparedness is more effective because there is a distrust of governmental authority. Many of the emergency managers either use or would like to use faith- and/or community-based organizations to reach these vulnerable populations. "Strictly speaking about emergency kits, there needs to be more community-based, faith-based churches that promote [emergency preparedness kits] and get them to their congregation and raise funds and get it to their community…" (IN EM 9).

Minority group leaders such as local representatives or church leaders may play a role in the dissemination of information to their constituents. One emergency manager faces challenges communicating emergency preparedness to a local Amish community. "We have reached out and met with Amish Bishops, they seem to be the people that the other Amish listen to…in some cases we have used Amish to talk to Amish. Some of the fire departments and emergency services have a high percentage of Amish members so we have reached out to that community to try to get them to have disaster preparedness kits" (PA EM 2). Another county with a large Somali population has used a local group and identified a community leader to arrange community meetings. The same community is also reaching out to the Hispanic population through community leaders. "We've actually talked about conducting CERT training or developing community emergency response team members in the Hispanic community. But at this point [it is] a big challenge [to find] someone who can do the interpreting for 40 hours of class" (MN EM 1). Sometimes, however, establishing these types of trusting relationships can take years. An emergency manager who communicates with a local American Indian reservation said that the local government and the group have had a relationship for 30 years, making the group more receptive to the idea of preparedness (ND EM 4).

Translators

Translators from minority communities and/or trained translators working for the local government are necessary in face-to-face communication. Many communities rely heavily on bilingual first responders. "The fire department and our police department have taken classes in Spanish and basically that's who people are going to deal with..." (IN EM 3). Another manager uses local university resources. "[County] has contacts at local universities including [university] who are fluent in a variety of languages. [Emergency manager] keeps a confidential file of bilingual emergency workers, such as the Chinese speaking police officer and the Japanese speaking firefighter. There are several Spanish speaking town employees who are CERT volunteers also" (CT EM 1).

Emergency managers spoke about the use of translators more in relation to crisis response than preparedness. Three managers specifically mentioned the use of Language Line. "We do subscribe to the AT&T Language Line that if people do call in where there is a language barrier that we can get an interpreter on the phone...that is for an emergency situation or a non-emergency situation where people may have questions or comments to make" (PA EM 2). Another emergency manager specified that Language Line was available only in emergencies (OK EM 1).

Communication Barriers

Subthemes of communication barriers identified in the transcripts included distrust of authority, limited ability to prepare, lack of perceived need to prepare, lack of personal responsibility, and emergency managers' limited budget and ability to communicate.

Distrust of Authority

The emergency managers suggested many minority groups must have additional prompting or assurances from trusted sources because of distrust of government or law enforcement. "Many folks come from communities or places where civil rights are being oppressed so when they come into the United States, they're very leery of what their government is going to do…many of these folks are here illegally, so when a police officer shows up or somebody in uniform shows up…my concern is that they're just not going to show up because they don't want to be identified" (MN EM 1). Much of the distrust can be attributed to misunderstanding of the function of government and law enforcement as "[immigrants and non-native speakers] tend not to watch the news, tend to be lower socioeconomic status" (PA EM 1). Other emergency managers were unaware of the potential effects of using government officials to communicate with some vulnerable populations. "We are trying to identify areas of low income housing based on census data. That way, if there is an emergency we can send law enforcement to those areas to inform them of what is happening" (ND EM 5).

One emergency manager related frustration from a situation in which communicating with ethnic populations failed during a hurricane. When the hurricane threat matured, local officials printed materials and had uniformed officers go from door to door to tell residents to move their cars to higher ground to save them in the case of tidal flooding. The manager found that many residents did not move their vehicles because they did not have a representative from the local community to legitimize the threat. "The more diverse the culture, the more important it is to have someone go in behind us..." (VA EM 1). Also, there were no minority representatives to translate the message from local officials. The emergency manager suggested relationships with "legitimate authorities" such as religious and medical leaders in the community.

Limited Ability to Prepare

An observation from many community emergency managers was that not everyone can afford to prepare emergency kits. A barrier to reaching everyone with awareness campaigns was identified as "Simple economics—to store enough food for three days is tough when you're working paycheck to paycheck.... It comes down to money" (TX EM 4). The emergency managers acknowledged that the preparedness campaigns were really for those who are able to and can afford to prepare the kits for their families, and more needs to be done to reach vulnerable populations through emergency services. Previous research on the development of emergency kits noted that "If a family has to decide between buying a gallon of milk or a gallon of gas, the campaigns promoted in McDonald's and movie theaters are neither reaching nor relevant to lower income households" (Veil, 2008, p. 390). From some of the emergency managers' perspectives, communicating the importance of building a kit to those who they know cannot afford the supplies needed for a kit is wasted effort.

Many racial/ethnic minority groups face economic disadvantages that keep them from purchasing items for a preparedness kit. One community with a large refugee population explained that some groups struggle to afford basic necessities on a daily basis. "Refugees came from a location where basic needs weren't being met; here we have Target and Wal-Marts, and the refugees are getting into these same habits of going out and buying items that day; if someone has one dollar they will go out and spend it on milk—something they need today" (ND EM 1). Money is a definite factor in the equation. "It's income, language, and culture. We all have the same influences. Preparedness kits are built from disposable income. They're not essential regardless…there's no perception of need" (TX EM 4).

Lack of Perceived Need

Some emergency managers noted that the perception of need may be lacking within some cultural ideologies as well. In some cultures, to discuss a crisis is to invite it into your life. In others planning for a future event that may or may not happen is wasted effort. "There's cultural challenges, too. For example, the Indian culture does not accept the idea of preparedness. You let life happen" (TX EM 5). Thus emergency managers must understand and relate to vulnerable populations on the population's level to create an awareness of and need for preparedness kits.

Lack of Personal Responsibility

A few of the emergency managers interviewed did not find themselves responsible for educating vulnerable populations about preparedness, citing the responsibility of individuals to prepare. "We are trying to map the community in order to get to areas that we know need help. We cannot prepare for them but must help them prepare. People have to take initiative to act. We need to give them information" (ND EM 5). The same emergency manager said, "Translating info and learning about disabilities is a courtesy and not a necessity." Another representative said communication is a "two-way street," and while government agencies can provide information, the ethnic community has a responsibility to communicate their needs (IN EM 4).

Emergency Managers' Budget and Ability

While most of the barriers identified centered on vulnerable populations' inability to receive and follow emergency preparedness messages, many emergency managers recognized their own inability to communicate messages. Almost all emergency managers cited a lack of resources, both financial and time-related, as a reason they are not able to adequately address crisis preparedness. The interviews indicated that emergency managers are concerned with the cost of reaching these populations in relation to benefiting the community as a whole. "We have no budget, no money to promote the kits" (TX EM 2). When asked how to best to reach these vulnerable populations, one emergency manager responded, "I am stretched pretty thin the way it is for the stuff. Trying to get multilingual resources out there—I don't even know where you'd start..." (IN EM 1). Also, in many cases, the minority populations do not constitute a large enough portion of the population to address. "Minorities are overlooked because there are so few in the area" (ND EM 5). The process of reaching these populations can be daunting, especially if there is a diverse range of minority populations. "We could get involved with multicultural organizations, but there's just too much to do and not enough time to do it" (TX EM 5).

Emergency managers were also hesitant to commit to reaching out to vulnerable populations. One emergency manager suggested that establishing communication with some groups may set up a level of expectation that emergency managers will need to uphold in order to maintain the trust of the community. "We have to make sure that if we do that we can meet those expectations. So that means we need to have people here who are going to be able to interpret the information in their languages on a consistent basis; otherwise they'll say, 'That's another thing we can't rely on'" (MN EM 1).

Analysis

The results of the interviews indicate there are multiple factors affecting the communication of preparedness messages with vulnerable populations. Vulnerable populations tend to be underrepresented and have few advocates, pushing their concerns and issues out of the awareness of mainstream society. Because they are typically misunderstood, both verbally and socially, non-English speakers are a trapped, muted group, unlikely to know about or ask for help regarding emergency preparedness. While minority groups may have close ties to family and others who share their language and cultural background, there is often a lack of leadership and resources within these communities. In many cases, religious organizations tie these communities together; however, emergency preparedness education may not be priority.

Emergency managers may ignore or not give adequate attention to vulnerable populations because financial limits make such preparation unfeasible. Since many aspects of emergency preparedness include an expense, those living at or below the poverty line cannot afford to prepare. Many Americans hovering around the poverty line may barely have the financial means to pay the bills or keep food on the table, making crisis preparedness neither an option nor a priority. Typically, these groups do not have Internet access or televisions to view preparedness messages, reliable methods of transportation for evacuation, or cell phone services to call for help. While emergency managers may feel communicating preparedness messages to those who cannot afford to prepare is wasted effort, ignoring economically disadvantaged groups has proven to greatly complicate emergency response in a crisis. During Hurricane Katrina, for example, the unmet needs of the disadvantaged in the city of New Orleans increased the death toll and hurt government credibility on local, state, and national levels.

Addressing Barriers

Little is likely to change unless emergency managers reconsider the existing emphasis on self-preparation. A renewed vision might include distributing emergency kits in times of crisis. Such distribution could be limited to those with the greatest need. Although this approach contradicts the emphasis on self-protection that is standard in crisis planning messages, it also confronts the undeniable fact that a notable portion of the population literally cannot follow the planning recommendations. There is awareness on the part of some of the emergency managers that these populations may be neglected and in need of special attention. "For years there have been accusations that [city] is not cared for as well—where there are more Hispanics and blacks" (TX EM 2). Attention in the form of special meetings, outreach programs, and alternative response strategies need to be available to these communities.

Financial Barriers

When emergency managers were asked what their major obstacle was in communicating with vulnerable populations, the overwhelming response was money. Despite the profound importance of preparedness communication, the visible aspect of an emergency manager's job is the response. To allocate funds on communication that could be spent on response equipment does not make sense to emergency managers who view public education as only a small part of their overall job responsibilities (Veil et al., 2009). In fact, the average county emergency management communication budget for reaching all publics is only $287 (Fisher Liu, 2007, p. 333). This fact is both discouraging and unwise. Hurricane Katrina proffered the bitter lesson that a failure to invest in crisis planning is, in the end, a very expensive proposition. Costs related to preventable injury and illness could be avoided by having better response capabilities in place before a disaster strikes.

Cultural Barriers

Unique cultural heritage and background must be taken into consideration when addressing vulnerable populations. In many cases, English is not spoken because individuals are first generation immigrants; therefore, non-English speakers are likely to have strong ties to their native cultures. The value of preparedness varies among cultures, and many come from places or backgrounds where preparedness is completely unheard of. New Americans may be unaccustomed to the new risk environment. For instance, Mexican immigrants finding employment in Colorado may have never experienced blizzard conditions. Therefore, appropriate education about major risks and consideration for a lack of understanding of the concept of preparedness must be anticipated.

Language Barriers

The most common communication tactic used by the emergency managers in reaching non-English speaking populations was creating and distributing materials translated into Spanish or another language. Two problems emerge from this dependence on the simple translation of materials. First, the translation process frequently fails to take into account the more complex cultural distinctions of many underrepresented populations. The decision-making process and hierarchy of influence, for example, differ widely among cultures. Expecting a message designed for the typical American citizen with a reasonable income to translate clearly to underrepresented populations is excessively optimistic at best and culturally insensitive at worst. Translation problems contribute to the second difficulty—a lack of motivation to acquire and read the materials.

Accessibility

Most managers said that the translated information available is on a website or in government services centers in the community. While the information may be available in a native language, a person would either have to have Internet access or physically go to a place where the material is distributed in order to obtain the information. The information may not be accessible and therefore not utilized. As one emergency manager said, "We've never had a request for [alternative language] materials. We've received no feedback that this was an issue. We set out some materials in Spanish, but they're rarely taken" (OK EM 1). Unused Spanish or other language material does not prove that individuals are not interested in preparedness but that the information is not reaching the target population; therefore, additional educational steps must be taken.

Developing Partnerships

The emergency managers interviewed sought to utilize pre-established faith and community groups as well as mobilize select community leaders in an effort to educate minority groups. Taking into account all the experiences, strategies, and suggestions from the emergency managers, it is evident that the formation of networks and co-programming among various government agencies and local groups may increase the chance that preparedness messages are heard and acted upon. Ulmer, Sellnow, and Seeger (2007) suggest developing partnerships with organizations prior to a crisis to aid in post-crisis communication and recovery. Just as organizations need support in communicating during a crisis, local government responsible for preparedness needs assistance in communicating their preparedness message and fostering relationships that may prove even more beneficial in the event of a crisis. Such relationships create a network through which culturally sensitive communicators can interpret pre-crisis messages and communicate them responsibly to their constituencies.

Partnering with Local Universities

One of the questions in the interview protocol not addressed in this study concerned educational resources available to emergency managers through local universities and colleges. The interviewers wanted to know what ways students, faculty, and staff could increase awareness, both internally and externally, about at-home preparedness kits. In general, emergency managers said they could use assistance with both research and practical application. "[City 1] and [City 2] really need assessments on their counties. If the universities could provide us with data on how many people have kits and how many don't, that would really help us out a lot. It would help us get funding as well to work towards this" (TX EM 1). Emergency managers also said that local universities and colleges can be beneficial in other ways, such as using bilingual students to communicate information, using the college or university television or radio station to disseminate public service announcements (PSAs), and using students in the distribution and assembly of preparedness kits. Students and other volunteers may be invaluable assets when communicating a message from the government. "We don't have the manpower to go to faith-based organizations. Some areas are very anti-government. Students will have better luck getting into some of these homeowners associations. Students could focus on neighborhoods" (OK EM 1). College students, then, are another resource at limited cost that can help bridge the gap between agencies and vulnerable populations.

Recruiting Volunteers

Individuals from within an ethnic group or minority community may serve as the disseminators of clear messages from a reliable source. When asked how these vulnerable populations are being reached, four emergency managers mentioned the use of CERTs (Community Emergency Response Teams). Members of CERTs are citizens trained in search and rescue, first aid, CPR, fire compression, etc. They learn how to triage victims and act as first responders in the case of an emergency. CERT training can be utilized in a number of ways, as emergency managers could target bilingual volunteers from inside and outside minority communities. Also, CERT training in Spanish or other languages for trusted community leaders, such as priests and pastors or local representatives, may be an effective way to reach minority audiences with preparedness awareness messages. One emergency manager attempted to create CERTs among members of the Hispanic community yet faced challenges in getting an interpreter for the training (MN

EM 1). In cases such as these, however, other relationships such as with bilingual students from a local college or university could assist in the training and presentation.

Reaching Out to Agencies and Schools

Partnerships with other organizations such as social support services, local community centers, religious organizations, and even public schools can create forums for which preparedness can be discussed and information can be distributed. Incorporating this information into the daily lives of the target population may keep vulnerable populations from fearing messages from government officials. Also, educating grade school students who are bilingual may be an effective means for communicating with non-English speaking parents. Grade school students may be more culturally in tune with mainstream society and are therefore more likely to accept information. For example, they may experience fire or storm drills, accept the threat, and take home prevention methods, either intentionally or unintentionally, to their parents.

Conclusion

Hurricane Katrina exposed the emergency management system's failures in accounting for and preparing our most vulnerable citizens. While much research is being conducted on how to limit personal risk, only recently have scholars begun studying risk that is dependent upon the realization of a crisis. Previous research indicates a number of challenges in communicating with vulnerable populations. This study looked at how emergency managers communicate pre-crisis information with vulnerable populations. Specifically, the efforts of local emergency managers to communicate preparedness were assessed to determine current strategies for and barriers to communicating preparedness with vulnerable populations.

Communication strategies identified included efforts to reach vulnerable populations through communication tactics, reliance on supporting agencies, involvement of community groups and leaders, and having translators available in a crisis response. Barriers to communicating with vulnerable populations included distrust of authority, limited ability to prepare, lack of perceived need to prepare, lack of personal responsibility, and emergency managers' limited budget and ability to communicate.

This study suggests that in meeting the unique needs of vulnerable populations, emergency managers must take into account the economic, cultural, and communication barriers that separate these populations from mainstream society in order to present culturally relevant material from trusted sources. Budgetary, time, and personnel constraints can be circumvented through co-programming and the formation of networks between various government entities and groups in the local community. Emergency managers need to reach vulnerable populations with preparedness messages because these groups typically face more challenges before, during, and after a crisis. Thus, effective pre-crisis communication can lessen the effects of a crisis and make emergency response efforts more effective.

References

Alexander, D. (2002). *Principles of emergency planning and management.* NY: Oxford University Press.

Andrulis, D. P., Siddiqui, N. J., & Gantner, J. L. (2007). Preparing racially and ethnically diverse communities for public health emergencies. *Health Affairs*, *26*(5), 1269–1279.

Blanchard, J. C., Haywood, Y., Stein, B. D., Tanielian, T. L., Stoto, M., & Lurie, N. (2005). In their own words: Lessons learned from those exposed to anthrax. *American Journal of Public Health, 95*(3), 489–495.

Botan, C., & Penchalapadu, P. (2008). *Assessing the role of public communication in state emergency operations plans (EOPs).* Paper presented at the National Communication Association Conference, San Diego, CA.

Brown, K. (2007, September 25). Testimony before the Subcommittee on Oversight of the House Committee on Ways and Means. Transcript retrieved from http://www.redcross.org/wwwfiles/Documents/GovernmentRelations/file_cont7084_lang0_2893.pdf

Chen, A., Keith, V., Leong, K., Airriess, C., Li, W., & Chung, K. (2007). Hurricane Katrina: Prior trauma, poverty and health among Vietnamese-American survivors. *International Nursing Review*, *54*(4), 324–331.

Cherry, A., & Cherry, M. (1997). A middle class response to disaster: FEMA's policies and problems. *Journal of Social Service Research*, *23*(1), 71–87.

Coombs, W. T. (2007). *Ongoing crisis communication: Planning, managing, and responding.* (2nd ed.). Thousand Oaks, CA: Sage.

Council for Excellence in Government (CEG). (2006). *Introducing the public readiness index: A survey-based tool to measure the preparedness of individuals, families, and communities.* Retrieved November 2008 from the Citizen's Corps Survey at www.citizencorps.
gov/pdf/cp_surveysdbase_62007.pdf

Federal Emergency Management Agency. (2004). Are you ready?: An in-depth guide to citizen preparedness. Retrieved from http://www.fema.gov/areyouready/

Federal Emergency Management Agency. (2009). Post Katrina: What it takes to cut the bureaucracy and assure a more rapid response after a catastrophic disaster. Statement of Craig Fugate before the House Committee on Transportation and Infrastructure, Subcommittee of Economic Development, Public Buildings and Emergency Management, U.S. House of Representatives Washington, DC. Retrieved from http://www.fema.gov/ pdf/about/testimony/072709_fugate.pdf

Federal Emergency Management Agency. (n.d). Preparing for disaster for people with disabilities and other special needs. [Brochure]. Retrieved from http://www.fema.gov/library/ viewRecord.do?id=1442

Federal Emergency Management Agency Ready America. (n.d.). Get a kit. Retrieved from http://www.ready.gov/america/getakit/index.html

Fisher Liu, B. (2007). Communicating with Hispanics about crises: How counties produce and provide Spanish-language disaster information. *Public Relations Review, 33,* 330–333.

Institute of Medicine. (2002). *Speaking of health: Assessing health communication strategies for diverse populations.* Washington, DC: National Academy Press.

Lachlan, K. A., Burke, J., Spence, P. R., & Griffin, D. (2009). Risk perceptions, race, and Hurricane Katrina. *The Howard Journal of Communications, 20,* 295–309.

Lindell, M. K., & Perry, R. W. (2004). *Communicating environmental risk in multiethnic communities.* Thousand Oaks, CA: Sage.

Littlefield, R., Rowan, K. E., Veil, S. R., Kisselburgh, L., Beauchamp, K., Vidoloff, K., Dick, M., Russell-Loretz, T., Kim, I., Ruvarac, A., Wang, Q., Siriko-Hoang, T., Neff, B., Toles-Patkin, T., Troester, R. L., Hyder, S., Venette, S. J., & Sellnow, T. L., (2010). We tell people. It's up to them to be prepared: Public relations practices of local emergency mangers. In W. T. Coombs & S. J. Holladay (Eds.), *Handbook of crisis communication* (pp. 245–260). New York: Wiley-Blackwell.

Macro International, Inc. (MACRO). (2006). A post-Katrina assessment. Citizen Preparedness Review, 2. Retrieved from http://www.citizencorps.gov/downloads/pdf/ready/citizen_prep_review_issue_2.pdf

Messias, D., & Lacy, E. (2007). Katrina-related health concerns of Latino survivors and evacuees. *Journal of Health Care for the Poor and Underserved, 18*(2), 443–464.

Mufliz, B. (2006). In the eye of the storm: How the government and private response to Hurricane Katrina failed Latinos. Retrieved from http://www.fachc.org/pdf/mig_ncla_Katrina_Report_FINAL.pdf

Murray-Johnson, L., Witt, K., Liu, W., & Hubbel, A. P. (2001). Addressing cultural orientations in fear appeals: Promoting AIDS-protective behaviors among Mexican immigrant and African American adolescents and American and Taiwanese college students. *Journal of Health Communication, 6*, 335–358.

National Council of Asian Pacific Americans [NCAPA]. (2005, September 29). *Katrina and the Asian-American community congressional briefing*. Retrieved from http://aajc.schipul.net/files/Katrina_Invite%20_3_%20_2_.pdf

Parker, D. J., & Handmer, J. W. (1998). The role of unofficial flood warning systems. *Journal of Contingencies and Crisis Management, 6*(1), 45–60.

Peguero, A. A. (2006). Latino disaster vulnerability: The dissemination of hurricane mitigation information among Florida's homeowners. *Hispanic Journal of Behavioral Sciences, 28*(1), 5–22.

Quinn, S. C., Thomas, T., & McAllister, C. (2005). Postal workers' perspectives on communication during the anthrax attack. *Biosecurity and Bioterrorism: Biodefense Strategy, Practice, and Science, 3,* 207–215.

Reynolds, B., & Seeger, M. W. (2005). Crisis and emergency risk communication as an integrative model. *Journal of Health Communication, 10*, 43–55.

Seeger, M. W., & Reynolds, B. (2007). Crisis communication and the public health: Integrated approaches and new imperatives. In M. W. Seeger, T. L. Sellnow, & R. R. Ulmer (Eds.), *Crisis communication and the public health* (pp. 3–20). Cresskill, NJ: Hampton.

Sellnow, T. L., & Vidoloff, K. G. (2009, September). Getting communication right. *Food Technology, 63*(9), 40–45.

Shin, H., & Bruno, R. (2003). U.*S.* Census Bureau language use and English-speaking ability: 2000. Retrieved from http://www.census.gov/prod/2003pubs/c2kbr-29.pdf

Shiu-Thornton, S., Balabis, J., Senturia, K., Tamayo, A., & Oberle, M. (2007). Disaster preparedness for limited English proficient communities: Medical interpreters as cultural brokers and gatekeepers. *Public Health Reports, 122*(4), 466–471.

Spence, P., Lachlan, K., & Griffin, D. (2007). Crisis communication, race, and natural disasters. *Journal of Black Studies, 37*, 539–554.

Ulmer, R. R., Sellnow, T. L., & Seeger, M. W. (2007). *Effective crisis communication: Moving from crisis to opportunity*. Thousand Oaks, CA: Sage.

U.S. Department of Health and Human Services. (2000). Healthy people 2010, 2nd ed., with understanding and objectives for improving health. Washington, D.C.: U.S. Government Printing Office. Retrieved October 12, 2007, from http://www.healthypeople.gov/Document/pdf/Volume1/11HealthCom.pdf

Uscher-Pines, L., Hausman, A. J., Powell, S., DeMara, P., Heake, G., & Hagen, H. G., (2009). Disaster preparedness of households with special needs in southeastern Pennsylvania. *American Journal of Preventive Medicine, 37*(3), 227–230.

Veil, S. R. (2008). Civic responsibility in a risk democracy. *Public Relations Review, 34*(4), 387–391.

Veil, S. R., & Husted, R. (in press). Best practices as an assessment for crisis communication. *Journal of Communication Management.*

Veil, S. R., Littlefield, R. S., & Rowan, K. E. (2009). Dissemination as success: Local emergency management communication practices. *Public Relations Review, 35*, 449–451.

Veil, S. R., Reynolds, B., Sellnow, T. L., & Seeger, M. W. (2008). Crisis and emergency risk communication as a theoretical framework for research and practice. *Health Promotion Practice, 9*(4), 26S–34S.

Zakour, M., & Harrell, E. (2003). Access to disaster services: Social work interventions for vulnerable populations. *Journal of Social Service Research, 30*(2), 27–54.

Chapter 11

A Multicultural Perspective on Crisis Management: Shell and the Ogoni People of Nigeria

Bolanle A. Olaniran & David E. Williams

This chapter explores the role of culture in crisis management, especially as it relates to the need for crisis preparation, planning, and management. There are two general categories of crises identified regarding culture: (1) crises within a specific culture and (2) crises that transcend one culture or involve activities of practitioners in more than one cultural boundary. The emphasis is placed on the second type of crisis, and a case study of the Nigerian Ogoni and a multinational oil company, Shell Petroleum Development Company, is presented to illustrate the importance of culture in crisis management in the globalization era. In other words, the crisis management teams were from one culture, but the affected publics were from different cultures.

Introduction

One of the few certainties about crisis is that it will happen and often at the least expected time. Meanwhile, crisis is rather indiscriminant as to where it occurs. It occurs in both wealthy countries and less economically developed countries (LEDCs). For example, we have witnessed the Bhopal chemical disaster in India, the Exxon oil spill, the debacle with Enron, and other corporate accounting scandals in the United States. The literature on crisis management has explored many variables including the environment, stakeholders' responses, media roles, and public reactions, among others. Culture is a less recognized yet critical variable to consider in crisis management. While the role of culture and cultural differences is well recognized in public relations research (e.g., Grunig & White, 1992; Kruckeberg & Starck, 1988), the role of cross-cultural communication has been explored very little in crisis communication. The implication of this omission in crisis literature is that it creates a significant void in understanding how to manage multicultural issues as they influence effectiveness and efficiency of crisis management programs from both corporate and academic perspectives. This chapter will attempt to fill this void by making a case for attention to multicultural issues in an effective crisis management program. The next section offers the rationale for adopting a multicultural perspective in crisis management.

Multicultural Perspectives in Crisis Management

Despite the claim that crises are inevitable due to the law of probability (Lerbinger, 1997; Perrow, 1984), they are preventable and manageable (Olaniran & Williams, 2001; Williams & Olaniran, 1994). However, the approach chosen in crisis management and communication is often based on a company's systematic assessment of the situation, which often fails to methodically take into account the issue of culture.

There are two general categories of crises regarding culture. The first includes those crises within a specific culture. The second involves crises that transcend one culture or involve activities of practitioners in more than one cultural boundary. For example, crises such as Johnson & Johnson's Tylenol tampering, *Exxon Valdez*'s oil spill, and AT&T's power outage all happened within one country or cultural boundary. However, other crises—such as the Bhopal chemical disaster, the Ford-Firestone tire recall, the contaminated Coca-Cola in Europe, and the issue with Shell and the Nigerian Ogoni people—all transcended national boundaries. The crisis management teams were from one culture, but the affected publics were from different cultures.

The first category of crises involves what is known in literature as *intra-cultural*, while the second category involves *intercultural/cross-cultural differences* (Kim, 1988; Olaniran & Williams, 2008). Given the multiple cultures involved, there is a need for public relations practitioners to take into consideration the cultural differences if crisis management plans are to be effective. The issue of culture in crisis management is germane to effective crisis management. Furthermore, the literature in cultural communication acknowledges the importance of an understanding of culture as a significant factor in the effectiveness of a particular strategy. For example, idiosyncrasies of outcomes for similar management strategies in different contexts are often attributed to the failure to take into account the cultural factors (Gudykunst & Kim, 1999; Olaniran, 2001). It stands to reason that each crisis situation is potentially cultural. Similarly, we live in an environment where distance is constantly shrinking, as globalization becomes the norm (Olaniran, 2001). Many crisis situations are potentially cultural. Similarly, when crises occur, they often involve practitioners from one culture who find themselves thrust into a different culture where the old rules or training do not produce expected results. Because crisis involves culture, it is beneficial to understand culture and variations in cultures across crises contexts. To that end, the next section will examine the dimensions of cultural variability.

Dimensions of Cultural Variability

Dimensions of cultural variability, which establish criteria for describing different cultures, lend some credence to the need to utilize a multicultural perspective in crisis management. There are four dimensions of cultural variability; they consist of power distance, uncertainty avoidance, individualism-collectivism, and masculinity-femininity (Hofstede, 1980; 1983). These four categories result from data collected from 50 countries and three world regions (Hofstede, 1980; 1983). These dimensions have been verified through research for operationalizing cultural differences and their effects on uncertainty reduction in intercultural communication encounters (Gudykunst, Chua & Gray, 1987; Olaniran, 1996; Olaniran & Stewart, 1996).

Power distance (PD) is defined as "the extent to which the less powerful members of institutions and organizations accept that power is distributed unequally" (Hofstede & Bond, 1984, p. 418). *Uncertainty avoidance* (UA) is defined as "the extent to which people feel threatened by ambiguous situations and have created beliefs and institutions that try to avoid these" (Hofstede & Bond, 1984, p. 419). *Individualism-collectivism* (IDCV) plays on the fact that in *individualistic* cultures, "people are supposed to look after themselves and their family only," while in *collectivistic* cultures, "people belong to in-groups or collectivities which are supposed to look after them in exchange for loyalty" (Hofstede & Bond, 1984, p. 419). *Masculinity-femininity* (MSFM) suggests that in *masculine* countries, "dominant values in society are success, money and things," while *femininity* refers to countries "in which dominant values are caring for others and quality of life" (Hofstede & Bond, 1984, p. 419-420). Along with societal culture, each organization has its own culture, which is explained in the following section.

Corporate or Organizational Culture

The issue of culture and the need for a multicultural perspective in crisis communication are further compounded by the fact that every organization has its own culture. The culture, while derived to some extent from societal culture, is for the most part reflective of the unique philosophy embraced by management and other employees. Organizational culture embraces factors such as control and consensus when describing the identity of an organization. One must take into consideration the role of societal culture in crisis management, as the culture is notably different from the dominant or prevailing organizational culture.

Research notes that in multinational corporations (MNCs), parent organizational partners usually exert control and dominance on the subsidiary affiliates (e.g., Olaniran, 2001). With this control, the parent company sets managerial standards and expectation guidelines, which in essence dictate procedural standards for the affiliates in other countries. Olaniran (2001) alludes to four control strategies depicting the attitudes of most MNCs: (1) ethnocentrism, (2) polycentrism, (3) regiocentrism, and (4) geocentrism (see also Perlmutter & Heenan, 1974).

The ethnocentric strategy implies that MNCs use home-country culture (i.e., its nationals) to manage the affiliates and subsequently determine the organizational culture. Often, ethnocentric views guide crisis management efforts and communication strategies in the sense that they advocate a homogenous public relations approach—despite the presence of different cultural contexts. However, the peril of the drive towards homogenized culture is believed to be the creation of a "fragmented" and "malignant" world culture (Steingard & Fitzgibbons, 1995). For example, differences in language, beliefs, and other cultural features are candidates for rendering standardized organizational protocols fruitless and ineffective (Olaniran, 2007).

Conversely, a polycentric strategy involves using the affiliates' nationals to manage their own operation with the affiliates' cultures. However, the driving corporate culture often serves as the overriding factor, thus taking power away from nationals in positions of authority. Subsequently, the MNCs' culture becomes the controlling entity.

Both regiocentric and geocentric strategies suffer certain degrees of insensitivity as well. Regiocentric strategies involve using people from a common market or from similar areas to manage MNCs. The geocentric strategy, on the other hand, involves using nationals from various places without emphasis on a particular culture or location to manage MNCs. While these approaches are supposed to involve members in the decision-making process, at times they serve quite the contrary. The corporate culture, which has a bias toward headquarters or the parent company's culture, often prevails in times of critical decision-making such as crisis management—thus giving way to the ethnocentric tendencies it is supposed to prevent. In essence, the relational power displayed during communication may be decided by the normative expectations of each organizational culture. Organizational culture has been described as "rule-like consensual meanings" that guide the symbolic interpretation of meaning and its use (Putnam, 1983). For instance, there are normative guidelines that influence the symbols attached to and the selection and appropriateness of a communication medium (Sitkin, et al., 1992), which is critical in crisis management. A case in point is the criticism of Exxon during the *Valdez* crisis when its CEO failed to fly to the scene of the spill and conduct press conferences. The inaction was perceived by the media and the affected public as insensitive and apathetic.

Knapp (1984) presents forms of address as indicators of relational formality. Such displays can be seen in crisis management as well. For instance, an organization with a formal hierarchical structure and whose cultural *ethos* emphasizes *formality* in superior-subordinate interactions would reflect this ethos by waiting for the top hierarchy before decisions are made, even when lower level individuals close to the crisis situations are well equipped to act (Olaniran & Williams, 2001).

Coordination and control, often referred to as human resource management, is essential to MNCs and most globalized industries (e.g., Adler & Bartholomew, 1992; Konsynski & McFarlan, 1990). The emphasis on coordination focuses on directing organizational activities toward goal attainment. Control, on the other hand, determines an organizational culture, which drives the coordination efforts in MNCs (See Geringer & Herbert, 1989 for review).

Similarly, it is noted that high-context cultures and non-individualistic cultures are more consensus-seeking than low-context cultures (e.g., Gudykunst et al., 1987; Ho & Raman, 1991; Oetzel, 1998). In a consensus-building, decision-making environment, which prevails in high-context cultures, emphasis is devoted to a decision-making process that minimizes internal conflict and maximizes internal group harmony (Hiltz & Turoff, 1978).

Furthermore, the persistence of global economic integration due to intensified competition demands more equitable international joint ventures of the kind between American and Japanese auto manufacturers. Consequently, transnational organizational cultures are becoming the norm rather than acculturating subsidiary partners into a dominant partner's culture (Adler & Bartholomew, 1992; Baker, Sircar, & Schkade, 1998; Martinsons & Westwood, 1997). Training to coordinate MNCs' activities becomes more pressing, especially in crisis management. Given these cultural differences, it is helpful to examine a specific crisis scenario or incident that vividly illustrates the need for multicultural perspectives in crisis management programs. This illustration can help demonstrate to MNCs and other global corporations the importance of culture considerations in their crises programs.

The Case of Shell and the Nigerian Ogoni People

The Shell Petroleum Development Company (SPDC) operates a joint venture agreement with the Nigerian National Petroleum Corporation (NNPC) (Royal Dutch/Shell, 1993). Shell's Nigerian joint venture has a capacity to produce one million barrels per day, which creates a significant flow of revenue for Shell. Thus, the significance of the joint venture as a revenue source cannot be overstated.

Many of Shell's gas flares are situated very close to villages, sometimes within one hundred meters of Ogoni homes. The company flares at some sites have operated continuously around the clock for more than 30 years (Saro-Wiwa, 1992, July 28). Greenpeace indicates that local reports present evidence of the company's atrocities, suggesting that:

> Apart from physical destruction to plants around the flaring areas, thick soot is deposited on building roofs of neighboring villages. Whenever it rains, the soot is washed off and the black-ink like water running down the roofs is believed to contain chemicals, which adversely affect the fertility of the soil. (Dappa-Biriye, Briggs, Idoniboye-Obu, & Fubara, 1992)

Perhaps the most troubling aspects of oil drilling to the Ogoni and Nigerian Delta people are the gas flares. According to Ken Saro-Wiwa (1992, July 28), author and spokesperson of the Movement for the Survival of Ogoni People (MOSOP), an organization established to call attention to environmental destruction and human rights issues concerning oil drilling in the Delta region of the country:

> The most notorious action of both companies [Shell and Chevron] has been the flaring of gas, sometimes in the middle of villages, as in Dere, (Bomu oilfield) or very close to human habitation as in the Yorla and Korokoro oil fields in Ogoni. This action has destroyed wildlife, and plant life, poisoned the atmosphere and therefore the inhabitants in the surrounding areas and made the residents half-deaf and prone to respiratory diseases. Whenever it rains in Ogoni, all we have is acid

> rain, which further poisons watercourses, streams, creeks and agricultural land. (see, http://archive.greenpeace.org/~comms/ken/refer .html/26)

Concerning the close proximity of its gas flares to Ogoni homes, Shell argues that this is a result of communities expanding into the vicinity of oil operations. The company maintains that once this happens, gas flares are relocated away from populous areas (SPDC publication, 1993). But, the Ogonis maintain that Shell has never relocated a gas flare (Saro-Wiwa, 1992).

Furthermore, it is noted that Shell's high-pressure pipelines pass above ground through villages and crisscross over lands that are used for agricultural purposes, thus rendering the lands economically worthless. Many pipelines also pass very close to Ogoni homes. Richard Tookey, a spokesperson for Shell, responded that the company is unable to bury pipelines beneath the ground because the lands are swampy and polluted; thus, he claims that burying the pipes will create hazards and will further risk pipe fractures and spillages (Tookey, 1992, December 9). Also, Tookey argued that when pipelines are close to inhabited homes, effort is made to re-route the pipelines away from those homes. Notwithstanding, Ogoni people recounted that their land was not swampy, and Shell did not make any attempt to re-route pipelines away from homes (Hinman, 1993).

According to an independent record of Shell's spills from 1982 to 1992, a total of 1,626,000 gallons were spilled in the company's Nigerian operations in 27 separate incidents (Oil Spill Intelligence Report, 1992). Of the total number of spills recorded from Shell—a company operating in more than 100 countries—40% of the spills were in Nigeria alone (Oil Spill Intelligence Report, 1992). Despite this staggering figure, it is still very difficult to determine the exact figures because many of the spills go unreported—therefore making the need to call attention to the effects of pollution and damages in the region all the more important. Shell maintains that most of its oil spills are due to sabotage. However, it is widely known that most spills in the country occur due to malfunctions and corrosion of equipment (Hinman, 1993; Oyekan & Ministry of Petroleum, 1991). The Nigerian Ministry of Petroleum Resources recorded 2,676 spillage incidents between 1976 and 1990 with the biggest cause being equipment malfunction at 38%, followed by corrosion at 21% and sabotage at 3% (Oyekan & Ministry of Petroleum, 1991).

Another dispute with oil companies in general, but Shell in particular, is the issue of compensation for land. While no individual owns mineral rights on a property in Nigeria, the land nonetheless belongs to indigenes.

However, Shell has been accused of acquiring land without paying compensation or securing approval from the owners. It is reported that whenever Shell acquires land in the Delta, it only pays for the crops growing on the land but not the land itself (Dappa-Biriye et al., 1992). Often, Shell and other oil companies' land negotiations are with illiterate villagers who are without proper legal counsel to explain or protect their interests.

The climax to these injustices culminated in a judicial inquiry following the Umuechem disaster (Dappa-Biriye, et al., 1992), where 80 innocent villagers were massacred at the hands of the Mobile Police Force during a protest against Shell (Okafor, 1998). There were many riots, but none garnered more attention than the plot—believed to be a collusion between Shell and then-Nigerian dictator president Gen. Sanni Abacha to get rid of MOSOP and its leaders—that resulted in the death of four Ogoni chiefs. Thus, Ken Saro-Wiwa (MOSOP leader) was arrested on May 21, 1994. The government alleged that he had conspired to kill four Ogoni chiefs in a riot earlier that day. The chiefs, however, had all been friends of Saro-Wiwa's, and one was an in-law. It is clear that the Nigerian security forces killed these men (Osaghae, 1995; TED, 1997). On October 31, 1995, Saro-Wiwa and his co-defendants were found guilty and sentenced to death by a *military tribunal* (Center for Constitutional Rights, n.d.; Okafor, 1998). International outrage caused some leaders, such as British Prime Minister John Major, along with Nelson Mandela, to call for clemency for Saro-Wiwa and his colleagues. Despite such pressure, the Nigerian dictatorship hanged Saro-Wiwa and eight other Ogoni men on November 10, 1995. However, it appears that Saro-Wiwa's and his colleagues' executions actually did succeed in calling the world's attention to Shell and other MNC oil conglomerates' operations in the Nigeria Delta Region. Consequently, what ensued was an explosive crisis that Shell can no longer afford to dodge. First, it is important to look at cultural dimensions and how they might have influenced the Ogoni and Shell crisis management.

Cultural Dimensions and Crisis Management

Collectivistic vs. Individualistic Cultures

Collectivistic and individualistic cultures view money differently. For instance, individualistic cultures (e.g., Western cultures) put greater emphasis on material wealth and money in terms of acquisition and use, while they put less emphasis on relationships than collectivistic cultures. In essence, one could infer that Shell and other oil companies, along with the

Western organizational perspective, are predominantly focused on revenue and profitability. Therefore, it is not surprising that they would want to keep as much of the revenue by paying out as little as they can afford to get away with.

Alternatively, it is noted that the Ogoni people are impoverished and survive through an agrarian economy. The general cultural view in an agrarian economy in Nigeria and in the Ogoni region is based on collective community development and help-providing activity for self-subsistence (Olaniran & Hall, 2002). A central aspect of the community development and help provision cultural perspective is to be a *good neighbor*. By being a good neighbor, it is understood that individuals must do all they can to live in harmony with their neighbors. Part of harmonious living includes placing little value on money—to the point of having to let go of a legitimate debt and other face-saving activities to preserve friendship and relationships at all cost (Olaniran & Hall, 2002). One must realize that the idea of going to a magistrate and filing a lawsuit against one's neighbor is looked upon negatively, such that it is considered more neighborly to suffer in silence rather than to lose friends in the aftermath of a legal victory. Failure to comply with harmonious norms is faced with punishment or chastisement (Olaniran & Hall, 2002).

Implementing Culturally Appropriate Crisis Management Practices in the Delta Region

As illustrated above, Shell's operations in Nigeria have been challenged by local unrest and criticism from communities within the oil-producing areas and have drawn increased condemnation from abroad. Certainly, greed and profitability played significant roles in the way Shell and other oil companies typically handled oil spill crises. As in most other corporate crises, the role of culture in selected strategies cannot be over-emphasized.

It appeared Shell was not alone in the crisis. The Nigerian government had its own share of blame for failing to regulate oil companies' activities to some degree. It was also known that, at the time of the crisis, the Nigerian government was under a notorious military dictator: General Sanni Abacha. For instance, large amounts of money were generated from oil production in the Ogoni region. From 1970 to 1988, the Federal Government received a total of $183.1 billion from oil extracted from the Delta region (Dappa-Biriye et al., 1992). From the Ogoni region alone, it has been estimated that Shell extracted over $30 billion in oil (Saro-Wiwa, 1992). However, there is little-to-no community improvement in the region, although Shell maintains that it has operated a community assistance program for over 25 years (Tookey, 1992). It is estimated that the money spent on community assistance during the same period is only $200,000 (Saro-Wiwa, 1992), or just 0.000007% of the value of oil extracted. This raised the question: how can any company justify this percentage as a tangible community assistance program?

In essence, Ogoni people were—and had been—taken advantage of, and they kept silent with little-to-no significant retaliation against Shell and the government's injustices or against the poor treatment of the people, their land, and overall source of livelihood. Further, Shell violated the cultural norm and the responsibility of being a *good neighbor* to its host. But with the aid of outsiders and organizations such as Greenpeace and other human rights watch groups, the Ogonis put together programs to help them *make right* some of their many wrongs. Calling attention to their rights from legal standpoints, it appears the Ogoni people know they are entitled to some compensation. How they go about it (e.g., holding oil workers hostage) is debatable, but there is no doubt that the strategy forced oil companies to address the issue.

Power Distance

The notion of power distance, which accepts that power is unevenly distributed in a culture, may help interpret actions from Shell toward the Ogoni people. When comparing United States' culture with Nigerian culture, it has been shown that Nigeria is a relatively more power-distant culture than America (Hofstede, 1980; Olaniran & Roach, 1994; Olaniran & Stewart, 1996). Thus, the ability to challenge authority, such as the government or its representative (i.e., Shell), is almost nonexistent. The Ogoni community could have perceived that Shell was an extension of the government (i.e., given the joint venture between the two) and thus considered it inappropriate to challenge Shell or other oil companies and their activities through political means. This argument has some merit because the court system that could provide arbitration is also controlled by the government and is perceived to be self-serving—as its primary obligation is to protect government interests or employers. Consequently, the Ogoni people found themselves in a no-win situation when they tried to seek justice.

From a cultural perspective it is important to realize that oil companies and other MNCs serve as indirect change agents. They either directly or indirectly take their ideas into a different culture, and over a period of time those ideas begin to reshape the culture and community in which they operate. For instance, Nigeria is not the only country where Shell operates joint venture oil explorations. However, Shell conveniently exploits the Ogonis' culture and situation to its own benefit. That is, it appears that Shell conformed only when the cultural mandates seemed to benefit its overall goal and scheme but resisted when the mandates did not favor its operational goals. The following statements from a British Petroleum (BP) engineer illustrate this point: "I have explored for oil in Venezuela, I have explored for oil in Kuwait, I have never seen an oil-rich town as completely impoverished as Oloibiri" (*African Concord*, 1990, December 3).

Even Shell has admitted as much but maintains that it is beyond the scope of its business activities to influence the government (Brak, 1993). But it was never outside the scope of its business practices to convince the government to send military police to crush civil riots or collude with the government to allow it to import guns and ammunitions for its security guards, police, and military (Dow Jones, 1996; Duodu, 1996; Ostapiak, 2002). While arming security guards may not seem to be something out of the ordinary to people from the United States, in Nigeria—with the exception of army and police—-ordinary citizens are not legally allowed to carry handguns and pistols. Consequently, whenever demands for land clean-up, community development, and compensation were made, Shell's response was often a burden-shifting strategy, making claims that it is the responsibility of the Nigerian government to direct 3% of its oil revenue back to the communities where the oil is produced (Hinman, 1993). However, in a similar situation in the United States, *Exxon Valdez*—and, very recently, BP—assumed responsibility to clean up oil spillage and compensate individuals whose livelihood was impacted by the spillage. As a matter of fact, the total settlement package from both civil and criminal cases against Exxon totaled $1 billion (*New York Times*, October 13, 1992). What is particularly interesting about the settlement is that not only does the settlement in the United States include claims for property damages but also punitive damages. The same can be said for the BP Gulf spill, where the company had to set aside $20 billion to address damages (Thomma & Talev, 2010). However, it appears that after much protest from civil rights and environmental activists, Shell was only interested in providing restitution for property damages (*New York Times*, October 13, 1992). Shell, in its defense, claimed that Ogoni people were inflating prices for property value (e.g., fishermen's nets). In essence, it appears that foreign oil companies operating in developing countries attempt to follow rules when it is convenient for them or the rules are to their advantage. They understand their obligations in the developed countries or their home countries but fail to abide by similar and humane standards when dealing with less economically developed countries.

Crisis Preparation

Unfortunately for Shell and other oil companies, there came a sudden change in government direction. The dictatorship rule of General Abacha came to an end when he died abruptly, and a new regime with a more democratic rule and humane treatment came to power, which left Shell and other oil companies scrambling. All of a sudden oil companies could no longer count on government "protectionism" to excuse their corporate mishaps, and they were forced to address compensation for damages, environmental factors, and lawsuits they had been trying to avoid all along. Only this time, they had no choice but to address them at once instead of gradually. The frequency of oil spills and failures to negotiate with the Ogoni people in a collaborative manner all showed the lack of an adequate crisis management plan by the oil companies. The lack of a concrete crisis management plan shows a reactionary, rather than a proactive, approach to oil spills in Nigeria.

Along with harmonious norms and cultural practices, oil companies must be vigilant by being crisis prepared—meaning, they must put together a workable plan along with frequent drills in order to continue to update their plans. Equipment for spill cleaning needs to be in place at all times. The key is to maintain control. Oil companies in Nigeria can learn a lesson from the likes of BP's and Exxon's mistakes in this respect. Exxon suffered from a lack of control and the inadequacy of its crisis preparation plan when it depended on the wire service (media) for its information instead of using the media to get its stories across (Williams & Olaniran, 1994). The best thing oil companies can do is to get the citizens on their side as sympathetic advocates, such that the media would have problems trying to escalate the crises. Likewise, periodic drills and clear preparation will demonstrate to the Ogoni people a commitment to their well-being in a manner consistent with their cultural norms. This approach acknowledges that the media are important, but it may not be as effective as it can be in collectivistic cultures. Rather, establishing and building relationships with key stakeholders and potential crisis victims seem more appropriate.

The anticipatory model of crisis management (Olaniran & Williams, 2001) suggests the need for crisis management efforts to focus on preventing crises to start with. Oil companies can avoid some unnecessary spills by getting equipment up to code. Needed repair or replacement of old and corroded parts are a must. Oil companies requested $3.7 billion from the government for overdue work (Tostevin, 1998). They need to be more adamant in making their demands, and they need to stress to their government partners the consequences following a lack of repair (e.g., spill, explosion, and loss of lives). If this strategy fails they must be willing to put concern for the people ahead of profit, along with a willingness to shut down operations.

Crisis management plans should incorporate people (e.g., company officials, members of the public, etc.) into the organizations' key decision making. This is necessary because they are able to offer insights on the specific cultures. These individuals must occupy relatively key posts rather than being relegated to token duties. This approach would be valuable as these natives serve, in a way, as representatives of the culture in helping corporations build long-lasting and trusting relationships with the public. The natives can also help determine how different crisis action plans are likely to be received by people in the communities and, hence, can help organizations safeguard against oversights and waste of resources while preventing crises from escalating. In essence, this approach can help organizations significantly reduce the level of uncertainties that surround crises in different cultures. In addition, some of these individuals can also serve as liaisons for negotiating smooth resolutions with the public.

Toward a Proactive Response

A necessary component of any effective crisis management program involves a proactive response from the crisis management team. The proactive stance involves taking the initiative in mediating crises through public relations efforts—but not as we know it in the Western world or culture. For instance, public relations, which is a key component of corporate image reparation, involves efforts from companies to repair and also establish good working relationships with the media. Therefore, in the Western culture, such as the United States, it is often recommended that oil companies or crisis-laden organizations take the initiative in providing the media with immediate information about a crisis, specifically regarding the necessary steps to be taken in order to address the crisis and relieve the pain and suffering of affected stakeholders (see Fink, 1986, on working with the media during a crisis). On the contrary, the collectivistic nature of most less economically developed countries—in particular, Nigeria—suggests that oil companies structure their public relations efforts around the notion of harmonious living. Therefore, it is implied that any public relations efforts would have to start with the affected citizens instead of the media. Aside from humanitarian reasons, this approach connotes the integrity on the part of the organization to deal with the crisis or prevent escalation because most developing countries' media do not currently enjoy the same degree of freedom of the press enjoyed in the United States and European countries (Olaniran & Williams, 2008). The media outlets are predominantly controlled by the government, and, given the partnership between the government and the respective oil companies, it would be impractical for government-owned media to publish damaging information against the government or projects embraced by the government (i.e., conflict of interest). We do not argue that good public relations efforts with the media are unnecessary—only that they may have to come second to the efforts concerning citizens and stakeholders. Specifically, Ogoni people must have the assurance that the oil companies care about them enough to notify them first-hand and warn them about any potential dangers that may be imminent rather than having to read about them in a newspaper or see them on television. Embracing such a strategy will acknowledge the cultural norms of the community, reinforce trust, and strengthen the relationship among parties. Key to this strategy is developing relationships with community leaders, such as local chiefs and kings, who can help corporations get their messages across by mobilizing the various groups in communities.

Crisis Response Strategies

A key lesson for crisis management is that deception and burden-shifting, or denial strategies, are mere avoidance techniques that only hide problems temporarily—even in developing countries. In essence, attempts to utilize any of these strategies evade responsibility, which only provides temporary and ultimately unacceptable responses to the crisis (Benoit, 1997; Coombs, 1999). Therefore, the following presents recommendations from the vigilant model of crisis management for oil companies in Nigeria. With the new democratic government and increased international attention to the Delta region, multinational oil companies will have much greater pressure from within Nigeria and world-wide to ensure a more humane response to oil spills.

A better approach for oil companies' crisis management is public relations work that demonstrates honesty and truth in information dissemination. Such efforts will require oil companies to refrain from automatic use of the burden-shifting strategy that blames every incidence of spills on sabotage (Olaniran & Williams, 2008). To claim sabotage immediately angers stakeholders who are already displeased about the crisis situation, and it destroys chances for a collaborative end to the crisis. For example, Shell did not help its already tarnished image when different government officials repudiated the company's claim that the fire was caused by sabotage from the people. On one hand, Shell's claim of sabotage implies that the 1,063 people who lost their lives brought it upon themselves. On the other hand, Shell's own former deputy managing director, Mr. Omene, later absolved the community from allegations of responsibility for the deadly fire and suggested the need for the company to put in place a contingency plan that activates occasional drills (Ogwuda, 1998; Okafor, 1998). Omene went on to substantiate that the combinations of inflammable hydrocarbon vapor and heat would naturally cause fire and explosion (Ogwuda, 1998; Okafor, 1998). This information is significant from the collectivistic cultural standpoint, where the emphasis is not so much on what is said or printed but rather on the nonverbal implication of what is said and, more specifically, how it is said. When one looks carefully at Shell's argument, the initial placement of the blame on the Ogoni people is strategically flawed and doomed to fail in any society, let alone in a society that places less emphasis on verbal and written communication. At the same time, Shell's argument damages whatever relationship may be there. Furthermore, once a relationship is damaged, it takes longer to repair, especially in a collectivistic culture. The admission of possible guilt by Shell did not do much to alleviate the problem, even if it was a strategic and calculated move of releasing truth to the public at a later time. In a collectivistic culture, it appears to be a very bad move, as it confirms what the Ogoni people have been claiming all along: any caring company with the people's interest at heart would not have endangered or lied about its responsibility to start with. A full apology (Coombs, 1999) would likely be a more effective response in this case.

Oil companies must also abandon their widely used denial and evasion of responsibility strategies (Benoit, 1997). Too frequently, oil companies have claimed that their obligations are fulfilled with tax payments and royalties to the government (*New York Times*, 1998). The assumption among the Ogoni people is that oil production—rather than regional development—has been the primary focus of the oil companies in the region. The worst image the oil companies could create comes from attempts to shy away from the region's *impoverishment* by being content with simply extracting oil and getting their money out (Olaniran & Williams, 2008). Although the approach in recent years seems to be changing, it is believed to be too little and too late

(Olaniran & Williams, 2008; *New York Times*, September 24, 1998). Notwithstanding, the company will have to continue improvements and take an active role in community development activities to improve the region and its relationships. It is the correct and ethical thing to do. Thus, it is hard to agree with Mr. Aramabi of Shell who claims that the government should take the lead in community development (Ibeanu, 2000; Onishi, 1998). Even then, this does not preclude Shell from community development activities of its own. After all, most companies engage in similar activities in their home countries (Connors, 1997). It is Shell's corporate obligation to do whatever it can to communicate to the community that it cares and to convince the people that it is doing all it can, even though it cannot do everything. It should not have to wait on the government. If Shell is doing its share and the community recognizes that fact in the final analysis, Shell will be spared in the grievance with the government. Such development activities are imperative because they will create peace and environmentally stable conditions, which have been absent in the region (Ogwuda, 1998).

Successful implementation of such a strategy demands communication and action that foster greater transparency in the relationship between government, community, and oil companies to alleviate crisis and suspicion. At no time should the companies make communities feel that they are running away from responsibilities or colluding with the government—and, more importantly, that they cannot be trusted. These companies are operating in a culture where citizens do not trust the government. Any perception of alignment or collusion with them at the expense of the people is bound to be met with suspicion and resistance. The major implication for crisis management is that trust is a key ingredient to any successful crisis preparation and management. One of the worst blunders MNCs can make is to get tangled in the host country's politics. Oil companies in Nigeria have ironically done just that by getting tangled in the web of Nigerian politics directly and indirectly by asking the military government for help instead of negotiating with the Delta people. This act backfired, even though the military successfully repressed the people by force (Connors, 1997). On the other hand, the action was a moral victory for the people because the incident got international attention. As a result, the people continue to use the same strategy—and rather frequently—to demand from the government fair allocation of oil revenues and increased participation in the government (Ogwuda, 1998). The Delta citizens also continue using the strategy of issuing ultimatums (Mbachu, 1998, November 4; November 10) to the companies to express their dissatisfaction with lack of payment and inadequate compensation for their damages.

Diffusion of Innovation

Oil companies also need to avoid paying off government officials and other individuals. A Shell spokesperson indicated that the company tries to avoid paying off individuals holding their employees hostage because once such payments start they will continue (Reuters News, 1998). The MNCs should instead actively pursue other negotiation strategies like adopting environmental clean ups, community development activities, and quicker responses to the people's demands. This process can be facilitated by embarking on the use of opinion leaders (i.e., chiefs and elders). The diffusion of innovation literature is rife with such successful use of opinion leaders to help in the dissemination of information about ideas or products to potential targets (Rogers, 1995). Perhaps no culture is more suited for the use of this approach than power distant cultures (e.g., Nigeria), where villages and towns are subject to the local governance by their town chiefs. These chiefs can serve as the go-between for MNCs and the individuals who are better able to convince people in their respective communities to rally behind corporate initiatives for regional development plans or preventive strategies to avoid crises and disasters (Olaniran & Williams, 2008). Taking this approach will help strengthen relationships and create mutual trust between the parties.

Furthermore, oil companies need to take the initiative in advocating independent agencies or embrace the special tribunal set up to adjudicate claims and compensation by victims rather than reject them. It is unfair anywhere to have the wrongdoer be in charge of determining fair compensation for the aggrieved party. Even a former Shell executive and the UN rapporteur concurred with this assessment that only independent agents consisting of experts in the oil and gas industry can adequately assess environmental damages to the Ogoni land and specify fair compensation (Obari & Akhaine, 1998; Ogwuda, 1998).

Multinational organizations operating in developing countries should always embark on voluntary ethical or humane practices. They should never have to wait until crises develop to be ethical. The ethical practices must be comparable to those in their home countries. Thus, organizations should never take advantage of people's ignorance or lack of education as was done in this case. Lack of education or simplicity of lifestyle does not make them any less human than members of these giant oil companies from the Western countries. In the final analysis, the people and those communities will recognize their rights—or someone will recognize those rights for them—and they will fight back. It is unfortunate that Ken Saro-Wiwa and his colleagues, along with thousands of other Ogoni people, died in the quest for fundamental rights. Then again, Saro-Wiwo's death accomplished what he could not accomplish while alive and gave Shell a crisis that is going to take a long time to repair.

As globalization continues, so does the need to address culture in crisis management and communication. It is apparent that one size does not fit all. While some universal cultural ideals—such as caring—may be useful, these ideals will have to be contextually or locally interpreted and used in a manner that addresses the idiosyncrasies of different cultures. And a company that does the best job of understanding these cultural idiosyncrasies will be best positioned to handle multicultural crises in the most successful manner. Therefore, for communication scholars and public relations practitioners, understanding messages and meanings becomes very critical and even more so than in the past, for this is the only way that complexities created by culture in crisis management can be addressed.

References

Adler, N. J., & Bartholomew, S. (1992). Academic and professional communities of discourse: Generating knowledge on transnational human resource management. *Journal of International Business Studies, 23*, 551–569.

Baker, J., Sircar, S., & Schkade, L. (1998). Complex document search for decision making. *Information & Management, 34*, 243–250.

Benoit, W. L. (1997). Image repair discourse and crisis communication. *Public Relations Review, 23*, 177–186.

Brak, A. J. (1993, August 1). Group public affairs coordinator, SIPC, London, in a letter to Paul Gilding, Executive Director of Greenpeace International. Retrieved from http://archive.greenpeace.org/comms/ken/refer.html

Center for Constitutional Rights (n.d). The case against Shell. Retrieved December 15, 2001, from http://ccrjustice.org/learn-more/faqs/factsheet%3A-case-against-royal-dutch/shell

Cohen, R. (1998, September 20). High claims in spill betray depth of Nigerian poverty. Retrieved December 3, 2000, from http://nytimes.com

Connors, L. (1997, April). Environmental racism: Australia, Shell & Nigeria. *Social Alternatives, 16(*2), 4–6.

Coombs, W. T. (1999). *Ongoing crisis communication.* Thousand Oaks, CA: Sage.

Custom oil spill data: Shell's ten year spill record, (1992). Oil spill intelligence report. Cutter Information Corp. Retrieved December 3, 2000, from http://archive.greenpeace.org/~comms/ken/refer.html/37

Dappa-Biriye, H. J., Briggs, R. R, Idoniboye-Obu, B., & Fubara, D. (1992). The endangered environment of the Niger delta constraints and strategies, an NGO Memorandum of the Rivers, Chiefs and Peoples Conference, for the World Conference of Indigenous Peoples on Environment and Development and the United Nations Conference on Environment and Development, Rio de Janeiro, p. 2. Retrieved from http://archive.greenpeace.org/comms/ken/refer.html

Duodu, C. (1996, January 28). Shell admits importing guns for Nigerian police. *Observer.* Retrieved from http://www.cohdn.ca/news/1-1/6.html

Fink, S. (1986). *Crisis management: Planning for the inevitable.* New York, NY: American Management Association.

Geringer, J. M., & Herbert, L. (1989). Control and performance of international joint ventures. *Journal of International Business Studies, 20,* 235–254.

Grunig, J. E., & White, J. (1992). The effect of worldviews on public relations theory and practices. In J. E. Grunig (Ed.), *Excellence in public relations and communication management*, (pp. 31–64). Mahwah, NJ: Lawrence Erlbaum.

Gudykunst, W. B., & Kim, Y. Y. (1999). *Communicating with strangers: An approach to intercultural communication (*3rd ed.). New York, NY: McGraw-Hill.

Gudykunst, W. B., Chua, E., & Gray, A. J. (1987). Cultural dissimilarities and uncertainty reduction processes. In M. McLaughlin (Ed.), *Communication yearbook, 10* (pp. 457–469). Beverly Hills, CA: Sage.

Hiltz, S. R., & Turoff, M. (1978). *The network nation: Human communication via computer.* Boston, MA: Addison-Wesley.

Hinman, P. (1993, August 25). Greenpeace exposes Shell's pollution record. Retrieved from http://www.greenleft.org.au/1993/112/5579

Ho, T. H., & Raman, K. S. (1991). The effect of GDSS and elected leadership on small group meetings. *Journal of Management Information Systems, 8(2),* 109–133.

Hofstede, G. (1980). *Culture's consequences*. Beverly Hills, CA: Sage.

———. (1983). Dimensions of national cultures in fifty countries and three regions. In J. Deregkowski, S. Dziurawiec, & R. Annis (Eds.), *Explorations in cross-cultural psychology,* (pp. 335–355). Lisse, Netherlands: Swets & Zeitlinger.

Hofstede, G., & Bond, M. (1984). Hofstede's culture dimensions: An independent validation using Rokeach's value survey. *Journal of Cross-Cultural Psychology, 15*, 417–433.

Ibeanu, O. (2000). Oiling the friction: Environmental conflict management in the Niger Delta, Nigeria. *Environmental Change and Security Report (6),* 19–32.

Judge accepts Exxon pact, ending suits on Valdez spill. (1992, October 13). *The New York Times*, pp. A14.

Kim, Y. Y. (1988). *Communication and cross-cultural adaptation.* London, England: Multilingual Matters.

Knapp, M. (1984). *Interpersonal communication and human relationships.* Boston, MA: Allyn & Bacon.

Konsynski, B., & McFarlan, W. (1990, September–October). Information partnerships—shared data, shared scale. *Harvard Business Review, 68,* 14–120.

Kruckeberg, D. & Starck, K. (1988). *Public relations and community: A reconstructed theory.* New York, NY: Praeger.

Lerbinger, O. (1997). *The crisis manager: Facing risk and responsibility*. Mahwah, NJ: Lawrence Erlbaum Associates.

Martinsons, M., & Westwood, R. (1997). Management information systems in the Chinese business culture: An explanatory theory. *Information & Management, 35*, 215–228.

Mbachu, D. (1998, November 4). Youth shun leaders plea to end Nigeria oil closures. *Reuters News.* Retrieved from http://www.reuters.com/news

Money demands delay return of shut Shell Nigerian oil. (1998, November 10). *Reuters News.* Retrieved October 15, 2000, from http://www.reuters.com/news

Nigeria and Shell: Partners in progress. (1993). SPDC publication Retrieved July 22, 1995, from http://archive.greenpeace.org/comms/ken/refer.html

Nigeria petroleum pollution in Ogoni region. (1997, November, 1). TED case studies Retrieved Juy 10, 2005, from http://www1.american.edu/TED/ogoni.htm

Obari, J. O., & Akhaine, S. (1998, November 30). Sorabjee urges probe of environmental ills, lauds transition. *Guardian News*. Retrieved January 20, 1999 from http://www.afbis.com/vanguard/default.htm

Oetzel, J. G. (1998). Explaining individual communication processes in homogeneous and heterogeneous groups through individualism-collectivism and self construal. *Human Communication Research, 25,* 202–224.

Ogwuda, A. (1998, November 30). Fresh disaster looms in Jesse. *Guardian News.* Retrieved December 3, 2000, from http://www.afbis.com/vanguard/default.htm

Okafor, C. (1998, November 30). Former Shell director absolves Jesse people of sabotage. *Guardian News.* Retrieved January 14, 1999, from http://www.afbis.com/vanguard/default.htm

Olaniran, B. A. (2007). Culture and communication challenges in virtual workspaces. In K. St. Amant (Ed.), *Linguistic and cultural online communication issues in the global age* (pp. 79–92*)*. Hershey, PA: Information Science Reference (IGI Global).

———. (2001). The effects of computer-mediated communication on transculturalism. In V. Milhouse, M. Asante, & P. Nwosu (Eds.) *Transcultural realities* (pp. 83–105). Thousand Oaks, CA: Sage.

———. (1996). Social skills acquisition: A closer look at foreign students on college campuses and factors influencing their level of social difficulty in social situations. *Communication Studies, 22*, 72–88.

Olaniran, B. A., & Hall, B. J. (2002). Helping behaviors and community building: The African Yoruba Tribe as a case example. *Journal of Intercultural Communication Research, 31*, 233–261.

Olaniran, B. A. & Roach, K. D. (1994). Communication apprehension in Nigerian culture. *Communication Quarterly, 42,* 379–389.

Olaniran, B. A. & Stewart, R. A. (1996). Instructional practices and classroom communication apprehension: A cultural explanation. *Communication Reports, 9,* 193–203.

Olaniran, B. A., & Williams, D. E. (2008). Applying anticipatory and relational perspectives to the Nigerian delta region oil crisis. *Public Relations Review, 34*, 57–59.

Olaniran, B. A., & Williams, D. E. (2001). Anticipatory model of crisis management: A vigilant response to technological crises. In R. L. Heath (Ed.) *Handbook of public relations* (pp. 487–500). Thousand Oaks, CA: Sage.

Oloibiri: In Limbo. (1990, December 3). *African Concord* (pp. 29).

Onishi, N. (1998, November, 22). Nigeria combustible as South's oil enriches North. *The New York Times.* Retrieved December 3, 2000, from http://www.nytimes.com/1998/11/22/world/nigeria-combustible-as-south-s-oil-enriches-north.html?pagewanted=all&src=pm

Osaghae, E. E. (1995). The Ogoni uprising: Oil, politics, minority agitation, and the future of the Nigerian state. *African Affairs*, *94*, 325–44.

Ostapiak, M. (2002, April). Youth in action: The truth about Shell oil. *Socialist Action.* Retrieved August 8, 2003, from http://www.socialistaction.org/news/200204/youth.html

Oyekan, A. J. & Ministry of Petroleum. (1991). *The Nigerian experience in health, safety and environmental matters during oil and gas exploration and production operations*. Paper presented at the First International Conference on Health, Safety and Environment in Oil and Gas Exploration and Production. (pp. 74). The Hague, Netherlands.

Perlmutter, H., & Heenan, D. (1974). How multinational should your managers be? *Harvard Business Review, 52(6)*, 121–132.

Perrow, C. (1984). *Normal accidents: Living with high risk technologies*. New York, NY: Basic Books.

Putnam, L. (1983). The interpretive perspective: An alternative to functionalism. In L. Putnam & M. Pacanowsky (Eds.), *Communication and organization: An interpretive approach* (pp. 31–54). Beverly Hills, CA: Sage.

Rogers, E. M. (1995). *Diffusion of innovations.* New York, NY: Free Press.

Royal Dutch/Shell. (1993, July). *The Shell Review* (pp. 20). Retrieved October 15, 2000, from http://archive.greenpeace.org/comms/ken/refer.html

Saro-Wiwa, K. (1992, July 28). *Statement of the Ogoni People to the tenth session of the working group on indigenous populations,* Palais des Nations, Geneva. Retrieved January 10, 2001, from http://archive.greenpeace.org/comms/ken/refer.html

———. (1992, December 8). Shell in Ogoni and the Niger delta, Emiroaf. Retrieved from http://archive.greenpeace.org/comms/ken/refer.html

———, K. (1992). Genocide in Nigeria: The Ogoni tragedy. (pp. 81). Port Harcourt: Saros. Retrieved December 23, 1999, from http://archive.greenpeace.org/~comms/ken/refer.html/28

Shell imports arms into Nigeria. (1996, April). *COHDN News*. Retrieved from http://www.cohdn.ca/news/1-1/6.html

Sitkin, S. B., Sutcliffe, K. M., & Barrios-Choplin, J. R. (1992). A dual capacity model of communication media choice in organizations. *Human Communication Research, 18,* 563–598.

Steingard, D. S, & Fitzgibbons, D. E. (1995). Challenging the juggernaut of globalization: A manifesto for academic praxis. *Journal of Organizational Change Management, 8(4)*, 30–54.

The case against Shell. Retrieved from Center for Constitutional Rights, http://wiwavshell.org /the-case-against-shell/

Thomma, S., & Talev, M. (2010, June 17). BP agrees to set aside $20 billion fund for Gulf oil spill. *The Olympian.* Retrieved October 17, 2010, from http://www.theolympian.com/ 2010/06/16/1275880/bp-agrees-to-set-aside-20-billion.html

Tookey, R. W. (1992, December 9). Letter to Shelley Braithwaite, London Rainforest Action Group, Shell International Petroleum Company Limited. Retrieved June 23, 2005, from http://archive.greenpeace.org/comms/ken/ refer.html

Tostevin, M. (1998, November 27). Nigeria oil firms increase demands for cash. *Reuters News.* Retrieved March 10, 2000, from http://www.reuters.com/news

Williams, D. E., & Olaniran, B. A. (1994). Exxon's decision making flaws: The hypervigilant response to the Valdez grounding. *Public Relations Review, 20,* 5–18.

Chapter 12

China: Growing Economy and Growing Crisis Potential

Lan Ni & Jennifer Vardeman-Winter

The authors of this chapter tackle a major endeavor as they address the crisis planning potential and differences in China. They discuss a well-organized approach to examining the realities of crisis planning in China by examining perspectives from the macro, meso, and micro levels. Insights are provided, ranging from the political-economic system challenges to organizational culture concerns and even individual and group level communication concerns. At each stage along the way, implications are offered for those who must tackle the challenge of utilizing pre-crisis planning in China.

Introduction

China is still a developing country despite its rapid growth (International Monetary Fund, 2010). Just in the past few years, the country has suffered from almost all the major types of crises identified by Coombs (1999). These crises included natural disasters (e.g., SARS and the Sichuan earthquake); malevolence (e.g., multiple cases of killing innocent kindergarten children to release anger against society); human breakdowns (e.g., mining accidents); challenges (e.g., boycotts against major corporations; Google withdrawal); megadamage (e.g., infant formula safety issue); organizational misdeeds (e.g., employee suicide cases at Foxconn); and rumors (regarding natural disasters). These crises have caused immense concerns both in and outside the country.

As China continues to become a fast-growing economic and political global superpower, crises of various kinds have emerged that give us reason to examine the culture of development as a peril as well as a supportive environment for organizations inside and outside of China. The Chinese government has faced considerable criticisms for its responses to many of these crises as well as its lack of preventative measures before crises occur. In addition, crisis communication planning is an underdeveloped area in China. It remains one of the weakest links in Chinese corporate communication (Wang, 2008). Only a little over half (52.2%) of companies had a department or function specifically for crisis management or communication in 2006 (Wang & Goodman, 2006). And the percentage of companies that have not set up crisis communication function increased to 73.3% in 2008 (Wang, 2008). How can pre-crisis planning be done in a more effective way in the cultural context of China?

The purpose of this chapter is to explore a cultural model of pre-crisis planning in the context of China. We examine important factors at different levels (macro, meso, and micro) that might influence or predict effective pre-crisis planning. These factors are then discussed in the framework of general pre-crisis planning principles. We conclude with suggestions on the methods for studying pre-crisis planning.

Overview

In general, the current literature on crisis management in public relations has focused on the organizational level. In other words, the unit of analysis is usually an organization that has experienced or is likely to experience a crisis (or crises). The data are usually taken from either the organization itself or the publics (or potential publics) who are affected by the crisis (or crises). And the conclusions or guidelines are usually drawn on behalf of the organization, (i.e., what lessons the organization should learn from the current crisis and what the organization needs to do in the future to prevent similar crises from happening again) (see Coombs and colleagues, e.g., Coombs & Holladay, 1996, 2002; Hagan, 2007, etc.).

However, in many cases, it is insufficient to examine only organizational-level factors for crisis management in general and for pre-crisis planning in particular. It would be a mistake to think that one can rest assured that crises could be prevented or prepared for effectively just because an organization has had a pre-crisis plan in place. A broader context needs to be taken into consideration when we discuss pre-crisis planning. This is particularly relevant with many developing countries where the national infrastructure in all aspects is sorely lacking or needs improvement. In this case, any single organization working within any single country will not be able to guarantee effective pre-crisis management.

In addition, because of the advent of communication technology and global business growth, crises are gaining global attention and having multinational impact (e.g., economic downturn of 2008; 9/11, which evolved in current Afghanistan and Iraq wars; and multiple suicide cases of Chinese workers in Foxconn company leading to worldwide reconsideration of labor conditions, etc.). Molleda (2010), for example, explored a transnational crisis perspective in addressing cross-national conflict shifting. Therefore, cultural contexts are important to consider in pre-crisis planning, especially because many crises increasingly involve multiple cultures (and different cultures may approach crisis preparation and response differently). Thus, a cultural model of pre-crisis planning is important for public relations.

To explore this model of pre-crisis planning within a broader context, we draw from global public relations theory of generic principles and specific variables (L. Grunig, J. Grunig, & Verčič, 1998; Sriramesh & Verčič, 2003; Vercic, L. Grunig, & J. Grunig, 1996). This theory has been important in exploring how public relations should be practiced in different parts of the world. It specifies five factors that will influence the practice of public relations. The factors are political-economic system, media system, level of development, extent of activism, and culture. Sriramesh and Verčič (2003) later provided another framework for studying international public relations by collapsing the five variables into a three-factor framework: infrastructure, media environment, and societal culture.

While this global public relations framework has been developed and tested (e.g., Rhee, 2002), little has been done to apply this framework to the crisis management situation. In addition, except for the inclusion of culture at the organizational level, this framework mainly focuses on macro-level factors within any given country, which are by nature more difficult to change. Factors at other levels that are critical to pre-crisis planning are not discussed.

To fill the gaps in research discussed above, we aim to explore a model of pre-crisis planning in the framework of the five-factor global public relations theory, with a focus on culture. China represents an exemplary nation in which to explore this model.

Traditional Pre-Crisis Planning Principles

Although largely understudied in public relations research (Coombs, 2010), pre-crisis communication can be divided into two areas. The first is the *strategic approach of crisis diversion*, which employs an anticipatory, issues management perspective to learning about potential issues and preventing them before they turn into crises (Coombs, 2009, 2010; Heath & Palenchar, 2009; Olaniran & Williams, 2001). Issues management emphasizes monitoring all groups' opinions and attitudes toward an organization and strategically prioritizing issues when they arise in order to cost-efficiently divert resources where necessary (Heath & Palenchar, 2009). More importantly, strategic issues management encourages organizations to maintain healthy, dialogic relationships with the public, using socially responsible communication practices like transparency in communication, bringing members of a public into decision-making, and acting as a good neighbor in communities. Organizations should also understand, anticipate, and immediately remedy potential problems, particularly technological failures/distress (because managers tend to make many assumptions about the safety of technologies, yet these are where many crises originate), such that "technology must be scrutinized for places and conditions in which a minute degradation could escalate exponentially" (Olaniran & Williams, 2001, p. 492). Furthermore, in addition to these models for pre-crisis management of information, crisis prevention includes rhetorical strategies that consist of practitioners "selling crises to management, inoculation, vigilance, plan efficacy, and self-efficacy of stakeholders for a crisis response" (Coombs, 2010, p. 479, citing Wan & Pfau, 2004; Williams & Olaniran, 1998).

The second area of pre-crisis planning is the *tactical approach to crisis response*, which entails the preparation process of practitioners copiously detailing the techniques and messages that will be sent to various affected groups during a likely crisis (Coombs, 2009). Best practices techniques attempt to reduce uncertainty before and during a crisis, and they include creating a crisis response plan, practicing the plan, designating spokespeople, designing and updating checklists and decision guides, preparing management with media training, preparing materials (e.g., press releases, fact sheets), determining distribution channels, and ensuring that materials are understandable and usable among special populations (Fearn-Banks, 2001; Seeger, Sellnow, & Ulmer, 2001).

China as a Unique Case for Pre-Crisis Cultural Planning

We believe that some critical factors need to be taken into consideration with pre-crisis planning. Pre-crisis planning is affected not only by the five factors in Sriramesh and colleagues' global public relations framework, but also by meso- and micro-level factors. We begin with macro-level factors.

Macro-level Factors

Political-Economic System. To a great extent, the cultural dimensions of the political and economic system in China have played an important role in explaining why certain crises have not been prevented in the first place or mitigated better once they occur. Currently, China has a unique political-economic system: pre-modern and modern in a world quickly becoming post-modern. The country maintains an authoritarian, communist political system. At the same time, it adopts market socialism (socialist market economy), a variation of the capitalist system where most of the economic development is guided by the market but the government still maintains strong macro-level oversight and influence (Chen, 2007). This gradual economic reform allows for the coexistence of multiple forms of ownership, which is suggested to be more effective for countries in transition to a market economy (Guthrie, 2006). The country's economy is developing quickly and shows great potential, but it is one full of peril. During the reform era that started in 1978, the average economic growth is more than nine percent with about eight percent per capita growth in gross domestic product (GDP) (Lippit, 2005). However, the political change is taking place in a context of a "deliberate and selective process of streamlining, professionalization, democratization, and centralization of the Chinese state apparatus" (Pieke, 2004, p. 517). In the meantime, the legal infrastructure is weak (Luo, 2002) so that "people in power positions" are still playing a more important role in decision-making of all kinds (Chen, 2007).

Implications for Pre-crisis Planning. The current political and economic system in China has created certain crises, which indicates a significant level of inappropriate pre-crisis planning in general. The current economic reform and the "open-door" policy, or the introduction of multiple forms of ownership, has boosted economic development to a great extent. However, the "market fundamentalism" (Lippit, 2005, p. 446), or a belief in the market system as always desirable and an inability to grasp the persuasiveness of market failures, has driven "market reform to unnecessary extremes," and an overly aggressive developmental mode has brought about many problems (Lippit, 2005, p. 452). Given the introduction of private enterprises without a proper legal and policy infrastructure, a lot of these enterprises can easily get away with wrongdoing. The main priority in the country at the moment is economic growth, so little regard is given to environmental concerns, public well-being, and morality. The pursuit of economic growth at the expense of public welfare distribution has, for example, led China to be ranked 144th out of 191 countries in 2000 in terms of public health service (Lippit, p. 446). Also, numerous coal mining accidents have occurred in China as a result of unsafe targets for production, rapid job creation, over-emphasis on reaching high profits, and an overall negligence of work safety.

In addition, the system of incentives used to boost economic growth has led to corruption at different levels because local officials have the power to guide (and benefit from) resource allocation. At the moment, career advancement of local officials depends more on speeding up development projects than on improving income levels of residents or providing social welfare. Therefore, public well-being tends to be ignored (Lippit, 2005).

At the same time, political reform in the country is still lagging far behind. The tight control under the authoritarian political system may not fit such rapid economic development for very long. Wang's (2005) study suggested that currently the legitimacy of the Chinese government largely depends on the fast economic performance with which the public is in general satisfied. However, the same long-term socioeconomic development, if continued to be achieved at the expense of ignoring public well-being, is going to bring about a complex of publics who will "more and more question its performance-based legitimacy" (p. 169). The rise of "critical citizens" is likely to arrive very soon in China, "challenging the government to manage both economic development and public demands for more democratic government institutions" (p. 169).

In fact, China's government is already experiencing a serious governance crisis right now (Ding, 2009). Ding argued that although some Internet-related policies were initiated to increase transparency and accountability of the government, these policies were based on the government's need to increase its legitimacy rather than to serve the people. Other hurdles such as the government's weak political credibility, digital divide, and information disparities presented challenges to effective implementation of these policies.

As a result, the government lacks credibility in pre-crisis planning—and businesses often take advantage of, rather than respond cautiously to, that shortcoming. The public is more likely to look for the hidden meanings in any message sent from the government. And the government's pre-crisis warning or lack thereof is not trusted. Rumors are spread more quickly and believed more than official government messages (e.g., earthquake rumors). In such a situation, it is no longer sufficient to only require certain organizations to have pre-crisis planning. In this case, China's State Seismological Bureau was doing all it could to stop, or reduce, the rumors, but it failed. Given this lack of credibility, when real disasters hit, such organizational level pre-crisis planning will not ensure adequate attention or compliance from the public.

In other words, the first and foremost priority in pre-crisis planning is to set conditions and develop strategies to restore and enhance the credibility of the government. The country has to strike a balance between high economic growth and attention to public well-being. Fortunately, the government has now realized the importance of pre-crisis prevention or diversion (Coombs, 2010) and has started from the macro-level, regulating through policy making. At the moment, the operation of small-scale (and thus more accident-prone) mining operations is strictly regulated.

Only when citizens truly buy the idea that the government has started to demand that major organizations build a "harmonious society" can these citizens participate in any pre-crisis efforts from the government. Also, does "harmonious" presume the asymmetry of a compliant population or one well served by government and business? Worth noting, for instance, is the absence of meaningful non-profit and advocacy activist groups, which operate the crisis environments of other countries.

Level of Economic Development

Although the economy is fast developing, the current level of development in China is still low. The GDP growth is continuing, reaching $5,878 trillion in 2010 (Central Intelligence Agency, 2009a). However, with China's huge population, GDP per capita is still low, at $6,600 in 2010, ranking at 125 (Central Intelligence Agency, 2009b).

The economic reform in China over the last decades has widened the income gap and caused socioeconomic polarization in urban areas (Chen, 2002). Many employees of state-owned enterprises (SOEs) have experienced a decline in their standard of living. They have been laid off or are receiving reduced wages and benefits. This has been closely correlated with the "omnipresent and organizational corruption" committed by government-appointed SOE directors and other high level officials (Chen, 2002).

In addition, urbanization, an integral part of economic development, is causing some serious challenges. Feng, Kugler, and Zak (2002) pointed out that migration of the agricultural population to the urban areas intensified in China, with the percentage of rural population dropping from 87.5 in 1952 to 69.6 in 1999. Rural workers continued to move to cities in search of jobs, and 114 million found jobs (Lippit, 2005). However, they faced awful working conditions and sometimes discrimination.

At the same time, the income gap between urban and rural population has also widened (Johnson, 2002). As a result, the current supply and demand of labor is in an imbalanced condition. That many multinational companies come to China to look for cheap labor is certainly not helping. With inconsistent regulations and weak enforcement of laws, policies, and regulations, many employees are forced to stay with their current employers because of the lack of alternatives.

Implications for Pre-crisis Planning. Current economic development has led to an increase of crises, many of which are difficult to predict or control at the pre-crisis stage. First, the widening income gap has generated hostility among many people toward the rich and powerful, and in turn, to society as a whole. Since early 2010, there have been seven incidents of hostile killings of innocent kindergarten children. This happened because people who believed they were treated unfairly (e.g., being laid off without sufficient reasons) wanted to release their anger on society.

At the same time, the broader economic environment needs to be taken into consideration. For example, the current imbalance in the supply and demand for labor has caused people to willingly accept lower than average salaries and heavier workloads. This has in turn caused many health problems, both mental and physical. Multiple employees in Foxconn Company committed suicide because of such mental and physical health problems. According to an article from CNN (2010), "The problems could be due to the pressures that many migrant workers face in China. They tend to be young people from the countryside who have never been away from home, working long hours, and feeling alone and vulnerable." Therefore, more mental health institutions should be established to help ease the anger and anxiety of people, especially that of frustrated and marginalized citizens.

The country as a whole needs to cope with the widening income–wealth gap. On one hand, the government needs to keep boosting the economy—but not at the expense of workers. To do this, the government has to establish and enforce stricter regulations regarding employment as well as to continue to increase employee and social security benefits. On the other hand, more nationwide campaigns about mental health awareness are needed to help people cope with the feelings of loss, frustration, anger, and anxiety that are potential consequences of dramatic economic and societal changes.

Level of Activism

The government in China does not encourage demonstration of activism, especially when it will threaten its stability. The process of applying for assembly, petition, and protest is overly complicated, so citizens are deterred from participating.

However, this low level of activism is changing gradually. In fact, as the economy keeps growing, the public is also becoming more liberal and pro-democracy, as indicated from a survey on three most economically developed regions in China (Wang, Rees, & Andreosso-O'Callaghan, 2004). The report showed that a majority saw a closer link between government administration and the quality of their daily lives. They were more politically assertive and politically tolerant. Such a change in the political and cultural orientation will make it more difficult for the ruling government to maintain its authoritarian rule.

In fact, given the restriction on actual and open activist efforts, there is a growing trend of Internet activism. Granted, most Internet activism is still targeted at major corporations rather than at the government because the online media environment is still monitored, too (see the next section). Typical examples are protesting foreign countries' inappropriate portrayal of China (e.g., the Beijing Olympics) and boycotts of products or services because of inappropriate remarks or behaviors (e.g., boycotting Sharon Stone films as a result of her Karma remark after China's earthquake). However, more and more members of an angry public are using the Internet to protest against the government, too. Frequently major crises are triggered by messages sent to and circulated on the Internet about controversial issues. These crises then create huge pressure on the government and force the government to respond.

Implications for Pre-crisis Planning. As a result, the level of activism is to be considered a key factor in pre-crisis planning. Both the Chinese government and other organizations (especially multinational companies) working in China have to pay more attention to public opinion than before, as the public is more empowered by information technology. Pre-crisis planning has to include constant monitoring of public opinion both off and on the Internet.

Media Environment

Closely related to the authoritarian political system, the media system in China is also tightly controlled. The traditional offline media are still closely monitored, and messages are screened by government officials. This can lead to ineffective crisis prevention of endemic diseases or disasters. For example, in a study on framing Severe Acute Respiratory Syndrome (SARS), Hong (2007) found that content on the Chinese Internet may have been different from that of the offline media but was still generally in compliance with the government's directives (i.e., not using much "attributing responsibility" frame in the case of SARS). Therefore, the Internet in China still does not serve the purpose of promoting open dialogue or rapid information dissemination in the face of infectious diseases.

However, even though China's government is still enforcing strict regulations on the Internet and restricting access to and from outside media, such information control is increasingly difficult to maintain. With more and faster technological advances, for example, the public is now able to get access to overseas media via certain proxy servers. This might influence how access to and dissemination of information online will change in the future. As a result, pre-crisis planning and prevention will have to be more open and transparent because with the advancement in technology, it is increasingly difficult to hide anything.

Societal Culture

Hofstede's (2001) cultural dimensions have traditionally been sufficient and effective in measuring societal culture. However, China is experiencing such a dramatic change that it is no longer adequate to only rely on those indexes.

Defining Culture. There are many definitions of culture. Kroeber and Kluckhohn (1952) defined culture as the patterns of human groups "acquired and transmitted by symbol," and the essential core of culture consisted of "traditional ideas and especially their attached values" (p. 180). Culture is also "the collective programming of the mind that distinguishes the members of one group or category of people from another" (Hofstede, 2001, p. 9).

Dimensions of Culture. The most comprehensive framework for cultural dimension to-date was developed by Hofstede (2001). It includes five dimensions, among which power distance, uncertainty avoidance, and individualism–collectivism seem to be directly relevant to employee relations. Thus, these three dimensions are discussed below.

Power distance is "the extent to which the less powerful members of institutions and organizations within a country expect and accept that power is distributed unequally" (Hofstede, 2001, p. 98). Gudykunst and Matsumoto (1996) pointed out that people in a high power distance culture do not question superiors' orders, whereas those in low power distance cultures do. In other words, people in a low power distance culture have a stronger desire to be informed, and the organizations' decisions need to be justified. Applied to employee relations, organizations have an obligation to disclose more to their employees.

Uncertainty avoidance is "the extent to which the members of a culture feel threatened by uncertain or unknown situations" (Hofstede, 2001, p. 161). This dimension is generally related to people's tolerance of ambiguity. Hofstede thought that members of a high uncertainty avoidance culture tend to resist changes and have higher levels of intolerance for ambiguity. Gudykunst and Matsumoto (1996) argued that in cultures high in uncertainty avoidance, a strong desire for consensus exists and deviant behavior is not acceptable.

The third dimension consists of two poles: individualism and collectivism. *Individualism* "stands for a society in which the ties between individuals are loose," whereas with collectivism, "people from birth onwards are integrated into strong, cohesive in-groups" (Hofstede, 2001, p. 225). This dimension states that in an individualistic culture, the individuals' goals are emphasized more than groups' goals. On the other hand, in a collectivistic culture, there will be more emphasis on in-groups. This definition shows that the welfare of people close to a group is important to maintaining a relationship with this group.

Chinese Culture beyond Cultural Dimensions. To explore the relationship between societal culture and relationship cultivation, the specific culture in China is also considered. Hung (2004) provided a comprehensive review of the literature on Chinese culture, which went beyond the more generic cultural dimensions offered by Hofstede (2001).

Hung's (2004) discussion of Chinese culture includes social orientations, such as family orientation and relational orientation. According to Yang (1992) and Triandis (1988), Chinese people consider "family" to be the basic unit in the society. The concept of self-OTHER indicates that in such collectivist cultures as China, people focus on a "we" identity and have a clear distinction between the "insider" and the "outsider."

Hung (2004) also used Yang's (1992) framework in discussing the relational orientation of Chinese people. We focused on those aspects that have a particular influence on how organizations handle relationships with employees: 1) role formalization, which means strict hierarchy in interpersonal relationships; 2) relational interdependence, which means each person defines his or her role by the interaction with other people; 3) relational reciprocity, the balance of the gives and takes in a relationship, which leads to the concepts of Ren Qing (favor) and Mian Zi (face) in regulating people's obligations to return favor and maintain a harmonious relationship where no one loses face; and 4) relational fatalism, which is used to explain why some relationships, either good or bad, exist.

Implications. Many cultural indicators have been undergoing dramatic changes in China. Whereas China (or regions sharing Chinese culture) used to be ranked high in collectivism, the young generation is more influenced by the western value system of individualism (Lau, 1992). Currently, collectivism is losing because the country no longer has the work unit (danwei) as an organizing or control system (Lau, 2001). The good old days are gone when the work unit used to take care of all the employees' health care, child care, housing, and even the employment of employees' children. Most employees no longer treat the work unit as the so called "big family."

Certainly, the focus on fame and fortune has made the country more masculine in nature. The pursuit of power and wealth has led to many short-term behaviors without considering the long-term consequences on the environment or on future generations.

With such dramatic changes in culture, Confucianism has lost the regulating power it once had in Chinese society. Other control systems have been changed (Lau, 2001). As a result, there has been a vacuum in belief systems and value systems. Lack of trust is commonplace. Many major food sources have been proved to contain harmful substances (e.g., eggs and infant formula), and the public has either become hypersensitive to food companies or has turned to fatalist thinking (i.e., since everything can be potentially harmful, I don't care about it anymore). The more pressing need in pre-crisis planning is thus to restore trust at all levels, that toward the society, the government (and government officials), and other people. Otherwise, all pre-crisis planning on the part of organizations or the government is not going to be received well.

Organizations should also pay attention to the unique cultural features in China. Organizations in China, whether from abroad or based domestically, should try to become the "in-groups" by taking care of their important publics. These publics will then consider them as a part of a collective and return with trust and good will. When crises hit, they can be more forgiving.

Meso-level Factors

At this level, pre-crisis planning can be influenced by the history and evolution of relationships between organizations and their public, organizational culture, and organizational-public power.

Relationship History and Evolution. Societal culture at the macro-level has an important bearing on the organizational level analysis of pre-crisis planning. China is a relationship-based society, and guanxi is predominant and deeply rooted.

Organizations, especially multinational corporations (MNCs), tend to have a complicated relationship with the public in China. On one hand, they provide more employment opportunities and help boost China's economy. On the other hand, these MNCs tend to exploit the cheap labor in China, not always paying attention to equal pay or benefits. However, many employee publics are bound by "barriers to exit" and cannot easily find an alternative. Therefore, they can be hostile toward these MNCs whenever certain crises hit.

Implications. Organizations need to use relationships as strategic resources to achieve organizational goals (Ni, 2006; 2009). Relationships built for strategic purposes will serve as a good foundation or buffer against future crises. Any pre-crisis planning needs to start with an assessment of the current relationship history and its evolution.

If organizations can pay more attention to their relational history with various groups, it may help to plan for crisis. For example, anticipating and planning post-crisis management of earlier crises may lead to higher quality relationships between an organization and its publics, thus preventing or mitigating future similar crises. Relationships based on symmetrically aligned interests can set a solid foundation for pre-crisis planning (Coombs, 2000; Coombs & Holladay, 2001). Liu's (2007) study in Taiwan supported such a theoretical argument in a Chinese cultural context. A supermarket that experienced two consecutive crises paid close attention to the evolution of relationships with its publics between the two crises. The relationships that were repaired after the first crisis served as a good foundation for pre-crisis planning for the second crisis.

Organizational Culture

Defining Organizational Culture. Sriramesh, J. Grunig, and Buffington (1992) defined organizational culture as "the sum total of shared values, symbols, meanings, beliefs, assumptions, and expectations that organize and integrate a group of people who work together" (p. 591).

Types of Organizational Culture. In their excellence study, L. Grunig, J. Grunig, and Dozier (2002) found two major types of organizational culture. In *authoritarian cultures,* decision-making is centralized, departmental agendas conflict with each other, and employees believe they are given little flexibility in innovation and that senior managers are interested in them only as workers—not as whole people. Employees may express fear of their supervisors, and such cultures are generally closed and resistant to external ideas. In organizations with *participative cultures*, the common value is teamwork. Employees believe management values them as whole people, and the organizations are open to ideas from external and internal environment.

Implications. It is good to have comprehensive pre-crisis planning in place. However, without an organizational culture that values symmetry and openness, such pre-crisis planning will not work well, as Marra (1998) illustrated.

In the Chinese context, the Foxconn Group has an authoritarian culture. The managers can publicly insult lower managers or employees. They can even ask lower-level managers or employees to stand in the corner to think over what they did wrong. Various kinds of formal and informal humiliation have caused mental stress for employees. This, coupled with continuous demands for working overtime, has caused 12 suicides just since the beginning of 2010.

The perceived normalcy of events revealed the dominance of an organizational culture that focused on control and efficiency and not care and humanity. Louis Woo, special assistant to Terry Gou, the founder of Foxconn's parent company, confessed, "For a long period of time I think we were kind of blinded by our success….We were kind of caught by surprise" (ABC news, 2010).

Organization–Public Power. Differences in organizational and public power have been examined in public relations research (see Berger, 2005; Dozier & Lauzen, 2000; J. Grunig, 2006; Holtzhausen & Voto, 2002; Karlberg, 1996). However, in crisis communication research, little examination of power and culture has occurred. When we talk about power it can be conceptualized in four ways in relation to crisis communication and organizational–public contexts:

1. *Power resource dependency theory* is how power flows through an environment and how it is distributed among players in an issue (Heath & Palenchar, 2009).
2. *Internal organizational power* is how power is distributed within an organization, particularly between public relations managers and technicians and the organizational dominant coalition. As the Excellence theory argued, if the public relations function is empowered by the dominant coalition, it can best serve its public through symmetrical communication (L. A. Grunig et al., 2002); thus, in a crisis situation, public relations managers would be empowered to conduct transparent, honest, meaningful, and useful communication with the public in need).
3. *External organizational power* is how organizations communicate to—or with—their public, particularly before, during, and after crises. Organizations have the choice to include stakeholders in their pre-, during- and post-crisis communication planning, and the public has the choice to oppose an organization's action (to some extent, depending on their perception of the problem, their level of involvement, and the constraints they have against communicating). Some scholars have argued that organizations always hold more power than a public because of the unequal balance of and access to resources between the two (Holtzhausen & Voto, 2002). To examine how power exists between organizations and their public, Berger (2005) offered a framework of power over relations (characterized by control and self-interest), power with relations (characterized by dialogue, negotiation, and shared power), and power to relations (characterized by resistance) (p. 6).
4. *Sociological, activist model* is how groups form movements themselves using the power resources they have—like access to certain media, established relationships within the community, and collective passion for remedying a situation—to combat an environmental, social, or organizational issue (see Holtzhausen, 2007).

Implications. How a power functions within a society, an organization, and a cultural group influences how pre-crisis planning will occur. For example, in China the government controls media organizations' structures and editorial decisions. Thus, the extent to which citizens are informed about risks and preventative actions they could take to reduce their threat is largely contingent upon the government's willingness to allow the media to tell these stories. Similarly, as China has lent money to many nations, including the United States, China must leverage its position of financial power when lending nations need information or assistance in worldwide crises. Exploring how these three contexts intersect—power relations, cultural groups' communication, and pre-crisis planning—is important to enabling power structures that can be useful during crises as well as filling in gaps where power relations are problematic.

Micro-level Factors

Crisis managers should examine identities, public formation around a crisis situation, and public reactions to crises and crisis responses from organizations. Some scholars recently have argued that public identities and cultural beliefs (i.e., shared identities and/or meanings about a crisis-based situation) determine how members of a group will act in a crisis situation (Sha, 2006; Sriramesh, Moghan, & Wei, 2007). For example, based on a group's shared identity and meanings, to what extent will they tolerate preventative measures against a certain type of crisis? These crisis communication considerations represent characteristics about groups that may help communicators know the extent to which groups will be aware of the crisis situation, be involved with the situation and motivated to act according to the protective actions, and be constrained in carrying out the suggested actions. This section will explore some recent considerations of identity as antecedent to communication behavior, public formation around crisis, and publics' external manifestation of crises and responses to organizations involved in crisis.

Identities

In the related field of risk management,[1] communication managers have traditionally segmented the public as a group of individuals sharing a basic demographic that is relevant to the risk situation (Aldoory, 2009). For example, people living along coasts and ports are taught hurricane preparedness, whereas others living inland are not; women are persuaded to conduct breast self-exams to detect breast cancers, whereas men typically are not. When culture has been considered in risk and crisis communication, a

[1] Risk communication is related to crisis communication because risk represents potential threats to health and well-being, whereas crisis is "risk manifested" (Heath & Palenchar, 2009, p. 275). Thus, it can be argued that pre-crisis communication is similar to risk communication when a particular risk may lead to crisis levels (e.g., disaster preparedness, public health epidemics, human error, terrorism events, etc.).

public has usually been identified according to single demographics (Aldoory, 2009). Although these univariate efforts have dominated risk and pre-crisis communication scholarship and practice for many years, scholars and communicators are currently considering deeper conceptualizations of group identities. These more recent conceptualizations include the intersectional perspective, the ecological perspective, and the community perspective.

Intersectional Perspective. Rather than segmenting a public according to single shared demographics, an intersectional approach to pre-crisis communication examines how a group's multiple identities (e.g., race, gender, location, nationality status, age, sexual orientation, disability, socioeconomic status, etc.) combine in interactive ways to form a unique experience from other groups (Vardeman-Winter & Tindall, 2010). As an interdisciplinary approach to understanding how different groups face such threats as persistent health disparities, intersectionality scholarship asks questions like:

> What is the meaning of race, ethnicity, class, gender, sexuality, and other systems of inequality across the ideological, political, and economic domains of society in institutional structures and in individual lives?; How are these co-constructed systems of inequality simultaneously produced, reinforced, and transformed—over time, in different social locations, and in different institutional domains (e.g., health, education, economy, religion, polity, family)? (Weber & Parra-Medina, 2003, p. 184)

These questions are important in pre-crisis planning to communicate effectively with groups who consistently experience a lack of information and awareness because of their simultaneously restrictive identities and environments. For example, emergency preparedness managers anywhere, including China, spend significant time considering the unique communication needs of special populations like those who are disabled, elderly, have low or no literacy, and those who do not speak the native language(s). However, an intersectional approach would also consider what other identity and environmental factors make the pre-crisis planning experience more difficult, such as living among a low-income population that also relies solely on public transportation or living in a group setting (like a nursing home or a disabled group home). Learning how these identity and environmental factors combine together to create distinct barriers to communication can enable more realistic pre-crisis behavioral change messages.

Ecological Perspective. The ecological perspective considers the broader factors that create the context by which certain groups may experience a crisis differently than other groups (Aldoory, 2009). The ecological perspective considers factors like the physical or organizational environment where a crisis occurs; the political landscape and current laws and policies impacting how authoritative groups will act in a crisis; historic events that lead to the shaping of a crisis and potential stigmas or social positioning of certain groups; social hierarchies, stereotypes, or practices that stymied or encouraged groups to accept or reject crisis preventative behaviors; epidemiological factors in which some groups experience certain public health problems in disproportionate amounts than others; and economic factors that hinder or enable certain groups to manage crisis themselves. Considering these other factors in pre-crisis management can alert communicators to the potential problems the public is already responding to, which may indicate why they are resistant to pre-crisis measures. For example, an ecological analysis of Foxconn's employee suicide crisis considers the organizational culture employees experienced—expectations for high levels of productivity with no employee assistance program—as a possible trigger to why employees felt overwhelmed to the point of ending their lives (CNN, 2010). Similarly, in the United States, an ecological approach to the obesity epidemic is investigating the "obesogenic environment," in which social scientists study aspects like how many local farmers' markets are available versus convenience stories versus supermarkets for a given area as well as how many points of interaction people have in an average day with unhealthy foods like vending machines and candy jars in an office setting (Plotnikoff, Lightfoot, Barrett, Spinola, & Predy, 2008).

Community Perspective. Rather than examining how individuals come into contact with pre-crisis messages and behaviors, the community perspective examines a defined community's collective beliefs, resources, and actions. A community can consist of people related physically—in terms of their locality to one another—or people loosely associated with one another because of a shared interest (Heath & Ni, 2010; Springston & Lariscy, 2010). A community approach "has the potential of achieving goals that are beyond the reach of individuals and has some important advantages over efforts by state or national agencies/organizations that often lack the local knowledge and support that a community-level coalition can achieve" (Springston & Lariscy, 2010, p. 547). Finding a community's motivations and barriers to action about a potential crisis has been a significant constraint to effectively preventing or lowering a community's risk for a crisis, particularly because communicating organizations often do not participate in (or come from) the communities that they are trying to change (Lupton, 1994). Intervening organizations currently employ a range of formal and informal methods—such as open houses, community advisory boards, and deliberative workshops—to engage community members in dialogues so that pre-crisis interventions and campaigns are enacted in ways that are meaningful, useful, and sustainable within and by the communities themselves (McComas, 2010).

Implications. These micro-level approaches to intimately understanding the identities of a public in pre-crisis planning have significant yet somewhat uncertain implications about the public in China, an evolving nation. For example, the intersectional approach is suited to use in pre-crisis planning in China because China's population consists of 56 different ethnic groups. However, Han Chinese make up 91.96% of the population (China's population mix, 2009); therefore, although much of crisis communication research has been conducted among Western crises and populations, many of these findings may not be easily translated to a population that, while it boasts many different ethnicities, is still dominated by one ethnic group.

Furthermore, an ecological approach considers environmental factors like geography. China's arable land is only 14.86% of the total land (Central Intelligence Agency, 2010), making it a country with tremendous resource shortages. Many geographical regions (such as the Yunnan, Guizhou, Sichuan, Chongqing, Gansu, Shaanxi, Hunan, and Hubei regions) are disaster-prone due to extreme weather and booming construction of highways and railways (Li, 2010).

Finally, China is a more community-based nation than are the United States and Europe, where most of the communication management research has emerged. Thus, research should be conducted to discover how Chinese communities confront pre-crisis planning and what aspects of Chinese community building can be exported as best practices to more individual-based nations.

Individual- and Group-Level Communication Variables

In conjunction with recent approaches to crisis communication as well as understanding of a public's identities and how they impact their communication behavior, certain variables have gained recent attention in pre-crisis communication management research because they may serve as antecedents to crisis response behavior. These individual behavior variables are threat literacy, shared risk and involvement, and emotional and personal attributes. Although this research has not been conducted among Chinese publics per se, future research can explore how these communication behaviors vary among different Chinese groups.

Threat Literacy. Emerging from the related concepts of social marketing, media literacy, and health literacy, threat literacy is applied to pre-crisis communication management as an individual's ability to self-assess one's personal risk for and susceptibility to a health/safety threat. Although threat literacy is still a fairly new concept, it is noteworthy in that it encompasses previous understandings of risk and crisis variables like perceived severity, susceptibility, self-efficacy, etc., combined with concepts from media and health literacy. For example, the National Network of Libraries of Medicine (NNLM) states that health literacy skills are "evaluating information for credibility and quality, analyzing relative risks and benefits, calculating dosages, interpreting test results, or locating health information" (Glassman, for the NNLM, 2010). Threat literacy, then, would include knowledge of how to perform protective actions against crises, knowledge of and access to the information sources being suggested, and knowledge of how to self-assess personal risk and susceptibility to radiation exposure (Chervin, Vardeman, & Osuji, 2009).

Shared Risk and Involvement. A series of studies (Aldoory, Kim, & Tindall, 2010; Aldoory & Van Dyke, 2006; Vardeman & Aldoory, 2008) investigated how groups seek information about a risk or crisis based on the extent to which they perceive that the victim and the organizational spokesperson in the media coverage about the event shared in the reader's perceived risk about and involvement in the threat. When providing participants with a news story about a food terrorism crisis, the researchers manipulated the quotations by the news story's victim and spokesperson to determine if participants perceived higher problem recognition and level of involvement within the logics of situational theory of publics. Aldoory et al. (2010) found that perceived degree of shared experience between the victim and spokesperson may increase problem recognition and information gaining (information processing and information seeking) in a crisis scenario and that these perceptions of shared experience may be antecedent to information gaining. In focus groups, Aldoory and Van Dyke (2006) learned that participants were more likely to seek information about a health crisis and act according to the suggested behaviors if they felt that the spokesperson was "in the same boat" (p. 356) as they were, since the spokesperson's quote demonstrated shared involvement and fear about the food contamination. Vardeman and Aldoory (2008) further learned that as the primary food-buyers and food-preparers for families, mothers were both likely to perceive and act upon shared spokesperson involvement in a similar food bioterrorism crisis and likely to distrust and resist an organization's attempt to share in consumers' risk and involvement. The importance of source similarity and message sensitivity is also recognized by Heath, Lee, and Ni (2009).

Emotional and Personal Attributes. Emotional responses and cognitive appraisals to crises have been studied as a way to prepare appropriate instructing messages for potential crises (Jin, 2010; Jin, Park, & Len-Ríos, 2010; Turner, 2006). Connecting affective factors to conflict management has been made within a contingency perspective, directing organizational strategies according to common emotional responses from previous disasters (Jin & Cameron, 2007). Jin (2010) found that participants were more likely to feel angry and sad the more they felt the organization had high predictability (or, how well the organization could have anticipated the crisis would occur) and high control/power in a crisis situation (or, the extent to which the organization could have exerted control during the crisis or prevented the crisis situation). Also, drawing on findings that members of a public can vary in their emotional reactions to different organizational responses (scapegoat, excuse, justification, apology, and accommodation) based on the amount of perceived predictability and controllability the organization had with the crisis, Jin concluded that "different combinations of discrete emotional engagement can be used in the future to identify and segment strategic publics in crisis communication" (p. 545).

A related area of research is an individual's or group's general tolerance for or acceptance of crises situations. The modern approach to crisis communication has argued that crises are time-oriented events of chaos, disrupting the normal order of operations between an organization and its public (Tyler, 2005). However, as risk and crisis are relative because groups may be differently involved in the crisis due to proximity, relevance, severity, susceptibility, self-efficacy, etc., what is a crisis to one involved group may be an inconvenience to another involved group. Thus, understanding groups' different levels of tolerance and factors determining those levels of tolerance are important for future research in pre-crisis planning.

Implications. One area that has not been investigated is the extent to which cultural beliefs about disasters impact a public's decision to seek information in a pre-crisis situation. Some cultures may understand crises to be part of the living process, whereas others may feel significantly more control over potential crises. Learning these cultural beliefs about crises can help communicators plan for preventative and instructing information around potential crises.

Public Formation Around Crisis

Some crisis research has examined what factors are important to a public when it forms in reaction to perceived threats against its well-being. This section describes recent research that explores how groups have formed before, during, or after crises based on some shared experience, meanings, or identities (i.e., culture). Unfortunately, less is known about how groups form when they feel threatened but not in an immediate crisis because crisis research tends to conduct case studies on groups and organizations who have already experienced crises. However, some of these studies lend insight into how cultural groups have responded to crises and therefore lend some insight into how pre-crisis communication can be conducted with a public.

After crises, a public may emerge to expose issues that had persisted in a social or physical environment but that ultimately manifest after a crisis reveals the public's marginalization. Crises present opportunities for a public that otherwise has been neglected to speak on the persisting issue (Kim & Dutta, 2009; Waymer & Heath, 2007). Reasons for previous latency around an issue can be attributed to a variety of factors, such as lack of resources, lack of information, or lack of time to commit to resolving an issue. For example, Hurricane Katrina exposed the gross disregard during pre-crisis natural disaster planning for groups that did not have resources to evacuate or take adequate shelter. In their evaluation of the post-crisis rhetoric, Waymer and Heath (2007) assessed the disparate frames being presented in public discourse by the organization being held responsible for remedying the group's crises (i.e., the federal government) and the frames presented by leaders of the affected groups (senators from affected or involved states). Similarly, Kim and Dutta (2009) studied a grassroots effort by people in New Orleans—named the Common Ground Collective—who used their shared experiences of living in New Orleans pre- and post-Hurricane Katrina to rebuild the physical, economic, and social structures for residents. Common Ground Collective's mission is based on solidarity that emerges from self-reflexivity, resistance to problems of oppression like racism and structural violence, and reconstruction based on "self-respect, self-empowerment, and self-defense" (Kim & Dutta, 2009, p. 158). For a cultural model of pre-crisis planning, crises that reveal issues like these suggest to communicators and community organizers to prepare publics to lend their voices in the early stages of an issue, always include marginalized voices in decision-making and public response about the crisis, and use reflective, inclusive, and proactive frames to shape an organization's and activist group's response (p. 106).

Implications. Much more research needs to be done to examine how cultural groups form around issues before crises occur. The key to this type of research is to gather observational data to build theory around how cultural elements (e.g., language, symbols, dominant versus subaltern meanings, norms, key events, significant relationships, and uses for media) interact with pre-crisis planning strategies. Furthermore, more research needs to be conducted to learn how activist and grassroots groups form, organize, make meaning of, and communicate about issues before and after crises.

Methods of Conducting and Researching Pre-crisis Planning

There are significant challenges with conducting and researching pre-crisis planning, primarily because of the uncertainty communicators face in the nature and extent of crises.

Challenges in Conducting Pre-crisis Planning

A number of challenges exist in pre-crisis planning because of the number of uncertainties associated with trying to predict the nature, location, publics affected, and long-lasting impacts of a crisis. Limitations exist in crisis communication research—mainly that traditional pre-crisis tools like the crisis communication plan and tabletop exercise drills are limited in their scopes and therefore have limited usefulness in being able to aptly prepare communicators and publics for potential crises (Tyler, 2005). Furthermore, many individual risks never manifest into crises or epidemic-level proportions and thus are not framed and communicated about as crises (e.g., the number of breast cancer deaths is not considered a crisis, and accident-related injuries are not framed as public crises). Therefore, there is a level of certainty that can occur in the public segmentation process that is missing from pre-crisis management. Finally, the lack of control over pre-crisis communication is limitless, as we often do not know when, where, in what capacity, or to what extent a crisis will occur.

For example, if we were constructing a crisis communication plan for a potential radiological attack somewhere in China, under which of the following circumstances would we design the preventive message campaign, based on the situational and physical environmental assessment?

a. The attack could occur with advance, little, or no notice.

b. The attack could be intentional or unintentional.
c. The attack could occur in a suburban or urban area—two options that would indicate very different concentrations of people in the same amount of space.
d. The attack could consist of one or multiple events.
e. The size of the attack would be largely unknown and could change over the course of time.
f. The effects of the actual radiological element(s) may be unknown for some time after the actual attack.

Other assessments—such as the communication resource capacity of the organizing group, the target audiences, and the audience strategies—have to be considered for how to communicate before and during such an event. The number of factors and combination of factors to be considered by an organization are almost endless, which suggests the need for uncertainty management (Peterson & Thompson, 2009).

Scholars have also considered the traditional ideologies and politics driving crisis communication discourse (Heath & O'Hair, 2009). Crisis communication literature has largely reflected a managerial bias, meaning that scholars have mainly examined crises from the perspective of organizations rather than from affected public perspectives (Kim & Dutta, 2009), which scholars argued makes crisis communication research, strategy, and responses deficient in being able to appropriately aid those in need (Waymer & Heath, 2007, pp. 88–89). Similarly, Tyler (2005) critiqued modern crisis communication practices by highlighting common assumptions, such as crises disrupt the normal operation of an organization and are problems that need to be solved; crises exist mostly at organizational levels; crises can be prevented; there is a need to regain control of a crisis; that an organization's publics are unified and have a common opinion and experience; and the spokesperson around a crisis is trying to help members of a public by giving them information in the interest of the public. To introduce more reality into pre-crisis and crisis communication, Tyler suggested that managers take a more postmodern approach, which encourages managers to accept and embrace the inherent ambiguity and complexity of crises and focus efforts on communication aspects that can be controlled.

Finally, new variations of traditional audience segmentation models have emerged that may help communicators better know which groups will be useful or unfavorable after a crisis (Kim, Ni, & Sha, 2008).

Potential Methods to Study Pre-crisis Planning

Since it is nearly impossible to know when and where a crisis will occur, our research options are limited to pre-crisis preparation and post-crisis responses. A significant amount of research in crisis communication has experimented with ways publics would respond in given crises (using simulated crisis situations) as well as analyzed how organizations have attempted to improve their images in the eyes of stakeholders after crises have occurred.

Some suggestions for future research to uncover how cultural groups manage crises have been made. For example, researchers should expand the scope of pre-crisis communication literature to non-managerial contexts. Also, researchers should attend to the issues of structure and agency in the realm of crises, and they can do so by participating in the cultural group's activities (Kim & Dutta, 2009). To do this, McComas (2010) presented an overview of methods for engaging communities to create dialogues around common issues.

Pre-crisis planning research from a cultural perspective also requires mixed methods and mixed samples. Studies should be conducted in which different kinds of crises are examined from the same cultural group's perspective to learn how the group gives meaning to and communicates about different types of crises, particularly according to their perceived severity, susceptibility, self-efficacy, response efficacy, and cues to action. Conversely, analyses of different cultural groups' pre-crisis planning about a common issue is helpful for learning the nuances of cross-cultural meanings and communication behavior as well as where common ground exists between the groups. Finally, combining ethnographic and experimental research can lend keen insight into how groups are likely to act and why they are likely to act around particular crises.

References

ABC news (2010). *iPhone-maker rallies workers after China suicides.* Retrieved November 11, 2010, from http://abcnews.go.com/Technology/wireStory?id=11433485

Aldoory, L. (2009). The ecological perspective and other ways to (re)consider cultural factors in risk communication. In R. L. Heath & H. D. O'Hair (Eds.), *Handbook of risk and crisis communication* (pp. 227–246). New York: Routledge.

Aldoory, L. Kim, J-K., & Tindall, N. (2010). The influence of perceived shared risk in crisis communication: Elaborating the situational theory of publics. *Public Relations Review, 36,* 134–140.

Aldoory, L., & Van Dyke, M. (2006). The role of perceived "shared" involvement and information overload in understanding how audiences make meaning of news about bioterrorism. *Journalism & Mass Communication Quarterly, 83*, 346–361.

Berger, B. K. (2005). Power over, power with, and power to relations: Critical reflections on public relations, the dominant coalition, and activism. *Journal of Public Relations Research, 17,* 5–28.

Central Intelligence Agency. (2009a). *The world factbook: Field listing GDP.* Retrieved August 5, 2010, from https://www.cia.gov/library/publications/the-world-factbook/fields/2195.html?countryName=&countryCode=&regionCode=%3Ehttp://www.theolympian.com/2010/06/16/1275880/bp-agrees-to-set-aside-20-billion.html

———. (2009b). *The world factbook: Country comparison GDP per capita.* Retrieved August 5, 2010, from https://www.cia.gov/library/publications/the-world-factbook/rankorder/2004rank.html

———. (2010). *The world factbook: China.* Retrieved September 8, 2010, from https://www.cia.gov/library/publications/the-world-factbook/geos/ch.html

Chen, A. (2002). Socio-economic polarization and political corruption in China: A study of the correlation. *Journal of Communist Studies and Transition Politics, 18*(2), 53–74.

Chen, Y-R. (2007). The strategic management of government affairs in China: How multinational corporations in China interact with the Chinese government. *Journal of Public Relations Research, 19*, 283–306.

Chervin, D. D., Vardeman, J. E., & Osuji, T. A. (2009, February). *Communicating to the public during radiological emergencies: PREPARE communication map.* Public Health Preparedness Summit 2009, sponsored by the National Association of County and City Health Officials, San Diego, CA.

China's population mix (2009). Retrieved September 8, 2010, from http://www.china.org.cn/e-groups/shaoshu/mix.htm http://www.theolympian.com/2010/06/16/1275880/bp-agrees-to-set-aside-20-billion.html

CNN (2010). *Report: China factory gives 2nd round of raises following suicides*. Retrieved November 11, 2010, from http://articles.cnn.com/2010-10-07/world/china.foxconn.raises_1_china-labour-bulletin-geoffrey-crothall-beijing-suicide-research?_s=PM:WORLDhttp://www.theolympian.com/2010/06/16/1275880/bp-agrees-to-set-aside-20-billion.html

Coombs, W. T. (1999). *Ongoing crisis communication: Planning, managing, and responding.* Thousand Oaks, CA: Sage.

———. (2000). Crisis management: Advantages of a relational perspective. In J. A. Ledingham & S. D. Bruning (Eds.), *Public relations as relationship management: A relational approach to the study and practice of public relations* (pp. 73–93). Mahwah, NJ: Lawrence Erlbaum.

———. (2009). Conceptualizing crisis communication. In R. L. Heath & H. D. O'Hair (Eds.), *Handbook of risk and crisis communication* (pp. 99–118). New York: Routledge.

———. (2010). Crisis communication: A developing field. In R. L. Heath (Ed.), *The Sage Handbook of Public Relations* (pp. 477–488). Thousand Oaks, CA: Sage.

Coombs, W. T., & Holladay, S. J. (1996). Communication and attributions in a crisis: An experimental study of crisis communication. *Journal of Public Relations Research, 8,* 279–295.

———. (2001). An extended examination of the crisis situations: A fusion of the relational management and symbolic approaches. *Journal of Public Relations Research, 13,* 321–340.

———. (2002). Helping managers protect reputational assets: Initial tests of the situational crisis communication theory. *Management Communication Quarterly, 16*, 165–186.

Ding, S. (2009). Informing the masses and heeding public opinion: China's new internet-related policy initiatives to deal with its governance crisis. *Journal of Information Technology & Politics, 6*, 31–42.

Dozier, D., & Lauzen, M. (2000). Liberating the intellectual domain from the practice: Public relations, activism, and the role of the scholar. *Journal of Public Relations Research, 12*(1), 3–22.

Fearn-Banks, K. (2001). Crisis communication: A review of some best practices. In R. L. Heath (Ed.), *Handbook of Public Relations* (pp. 479–485). Thousand Oaks, CA: Sage.

Feng, Y., Kugler, J., & Zak, P. (2002). Population growth, urbanisation and the role of government in China: A political economic model of demographic change. *Urban Studies, 39,* 2329–2343.

Glassman, P. (2010, January 5). U.S. National Network of Libraries of Medicine: Health literacy. Retrieved August 23, 2010, from http://nnlm.gov/outreach/consumer/hlthlit.html#A2http://www.theolympian.com/2010/06/16/1275880/bp-agrees-to-set-aside-20-billion.html

Grunig, J. E. (2006). Furnishing the edifice: Ongoing research on public relations as a strategic management function. *Journal of Public Relations Research, 18,* 151–176.

Grunig, L. A., Grunig, J. E., & Dozier, D. M. (2002). *Excellent public relations and effective organizations: A study of communication management in three countries.* Mahwah, NJ: Lawrence Erlbaum.

Grunig, L. A., Grunig, J. E., & Verčič, D. (1998). Are the IABC's excellence principles generic? Comparing Slovenia and the United States, the United Kingdom, and Canada. *Journal of Communication Management, 2,* 335–356.

Gudykunst, W. B., & Matsumoto, Y. (1996). Cross-cultural variability of communication in personal relationships. In W. B. Gudykunst, S. Ting-Toomey, & T. Nishida (Eds.), *Communication in personal relationships across cultures* (pp. 19–56). Thousand Oaks, CA: Sage.

Guthrie, D. (2006). *China and globalization: The social, economic and political transformation of Chinese society*. New York: Routledge.

Hagan, L. M. (2007). For reputation's sake: Managing crisis communication. In E. L. Toth (Ed.), *The future of excellence in public relations and communication management: Challenges for the next generation* (pp. 413–440). Mahwah, NJ: Lawrence Erlbaum.

Heath, R. L., Lee, J., & Ni, L. (2009). Crisis and risk approaches to emergency management planning and communication: The role of similarity and sensitivity. *Journal of Public Relations Research, 21*, 123–141.

Heath, R. L., & Ni, L. (2010). Community relations and corporate social responsibility. In R. L. Heath (Ed.), *The Sage Handbook of Public Relations* (pp. 557–568). Thousand Oaks, CA: Sage.

Heath, R. L., & O'Hair, H. D. (2009). The significance of crisis and risk communication. In R. L. Heath & H. D. O'Hair (Eds.), *Handbook of risk and crisis communication* (pp. 5–30). New York: Routledge.

Heath, R. L., & Palenchar, M. J. (2009). *Strategic issues management: Organizations and public policy challenges* (2nd ed.). Thousand Oaks, CA: Sage.

Hofstede, G. (2001). *Culture's consequences: Comparing values, behaviors, institutions, and organizations across nations.* (2nd ed.). Thousand Oaks, CA: Sage.

Holtzhausen, D. R. (2007). Activism. In E. L. Toth (Ed.), *The future of excellence in public relations and communication management: Challenges for the next generation* (pp. 357–379). Mahwah, NJ: Lawrence Erlbaum.

Holtzhausen, D. R., & Voto, R. (2002). Resistance from the margins: The postmodern public relations practitioner as organizational activist. *Journal of Public Relations Research, 14,* 57–84.

Hong, T. (2007). Information control in time of crisis: The framing of SARS in China-based newspapers and internet sources. *Cyberpsychology and Behavior, 10*, 696–699.

Hung, C. J. F. (2004). Cultural influence on relationship cultivation strategies: Multinational companies in China. *Journal of Communication Management, 8*, 264–281.

International Monetary Fund (2010). Emerging and developing economies. Retrieved September 14, 2010, from http://www.imf.org/external/pubs/ft/weo/2010/01/weodata/groups. htm#oemhttp://www.theolympian.com/2010/06/16/1275880/bp-agrees-to-set-aside-20-billion.html

Jin, Y. (2010). Making sense sensibly in crisis communication: How publics' crisis appraisals influence their negative emotions, coping strategy preferences, and crisis response acceptance. *Communication Research, 37*, 522–552.

Jin, Y., & Cameron, G. T. (2007). The effects of threat type and duration on public relations practitioners' cognitive, affective, and conative responses to crisis situations. *Journal of Public Relations Research, 19*, 255–281.

Jin, Y., Park, S-A., & Len-Ríos, M. E. (2010). Strategic communication of hope and anger: A case of Duke University's conflict management with multiple publics. *Public Relations Review, 36*, 63–65.

Johnson, D. G. (2002). Can agricultural labour adjustment occur primarily through creation of rural non-farm jobs in China? *Urban Studies, 39*, 2163–2174.

Karlberg, M. (1996). Remembering the public in public relations research: From theoretical to operational symmetry. *Journal of Public Relations Research, 8*, 263–278.

Kim, I., & Dutta, M. J. (2009). Studying crisis communication from the subaltern studies framework: Grassroots activism in the wake of Hurricane Katrina. *Journal of Public Relations Research, 21,* 142–164.

Kim, J-N., Ni, L., & Sha, B-L. (2008). Breaking down the stakeholder environment: Explicating approaches to the segmentation of publics for public relations research. *Journalism & Mass Communication Quarterly, 85*, 751–768.

Kroeber, A. L., & Kluckhohn, C. (1952). Culture: A critical review of concepts and definitions. *Papers of the Peabody Museum of American Archeology and Ethnology, 47*, 41–72.

Lau, R. W. K. (2001). Socio-political control in urban China: Changes and crisis. *British Journal of Sociology, 52*, 605–620.

Lau, S. (1992). Collectivism's individualism: Value preference, personal control, and the desire for freedom among Chinese in Mainland China, Hong Kong, and Singapore. *Personality and Individual Difference, 13*, 361–366.

Li, P. (2010, August 10). China identifies 16,000 disaster-prone areas. *China.org.cn.* Retrieved November 15, 2010, from http://www.china.org.cn/china/2010-08/20/content_20757826.htmhttp://www.theolympian.com/2010/06/16/1275880/bp-agrees-to-set-aside-20-billion.html

Lippit, V. D. (2005). The political economy of China's economic reform: Observations on China and socialism. *Critical Asian Studies, 37*, 441–462.

Liu, T. (2007). The relationship between pre-crisis, crisis and post-crisis management: A case study of Watsons Inc. Taiwan. Unpublished master's thesis, University of Houston.

Luo, Y. (2002). *Multinational enterprises in emerging markets*. Copenhagen, Denmark: Copenhagen Business School Press.

Lupton, D. (1994). Toward the development of critical health communication praxis. *Health Communication, 6*, 55–67.

Marra, F. (1998). Crisis communication plans: Poor predictors of excellent crisis public relations. *Public Relations Review, 24,* 461–474.

McComas, K. A. (2010). Community engagement and risk communication. In R. L. Heath (Ed.), *The Sage Handbook of Public Relations* (pp. 461–476). Thousand Oaks, CA: Sage.

Molleda, J-C. (2010). Cross-national conflict shifting: A transnational crisis perspective in global public relations. In R. L. Heath (Ed.), *The Sage Handbook of Public Relations* (pp. 223–235). Thousand Oaks, CA: Sage.

Ni, L. (2006). Relationships as organizational resources: Examining public relations impact through its connection with organizational strategies. *Public Relations Review, 32,* 276–281.

———. (2009). Strategic role of relationship building: Perceived links between employee–organization relationships and globalization strategies. *Journal of Public Relations Research, 21,* 100–120.

Olaniran, B. A., & Williams, D. E. (2001). Anticipatory model of crisis management: A vigilant response to technological crisis. In R. L. Heath (Ed.), *Handbook of Public Relations* (pp. 487–500). Thousand Oaks, CA: Sage.

Peterson, T. R., & Thompson, J. L. (2009). Environmental risk communication: Responding to challenges of complexity and uncertainty. In R. L. Heath & H. D. O'Hair (Eds.), *Handbook of risk and crisis communication* (pp. 591–606). New York: Routledge.

Pieke, F. N. (2004). Contours of an anthropology of the Chinese state: Political structure, agency and economic development in rural China. *Journal of Royal Anthropological Institute, 10*, 517–538.

Plotnikoff, R. C., Lightfoot, P., Barrett, L., Spinola, C., & Predy, G. (2008). A framework for addressing the global obesity epidemic locally: The Child Health Ecological Surveillance System (CHESS). *Preventing Chronic Disease, 5*(3). Retrieved August 25, 2010, from http://www.cdc.gov/pcd/issues/2008/jul/07_0007.htm

Rhee, Y. (2002). Global public relations: A cross-cultural study of the excellence theory in South Korea. *Journal of Public Relations Research, 14*, 159–184.

Seeger, M. W., Sellnow, T. L., & Ulmer, R. R. (2001). Public relations and crisis communication: Organizing and chaos. In R. L. Heath (Ed.), *Handbook of Public Relations* (pp. 155–165). Thousand Oaks, CA: Sage.

Sha, B-L. (2006). Cultural identity in the segmentation of publics: An emerging theory of intercultural public relations. *Journal of Public Relations Research, 18,* 45–65.

Springston, J. K., & Lariscy, R. W. (2010). The role of public relations in promoting healthy communities. In R. L. Heath (Ed.), *The Sage Handbook of Public Relations* (pp. 547–556). Thousand Oaks, CA: Sage.

Sriramesh, K., Grunig, J. E., & Buffington, J. (1992). Corporate culture and public relations. In J. E. Grunig (Ed.), *Excellence in public relations and communication management* (pp. 577–595). Hillsdale, NJ: Lawrence Erlbaum.

Sriramesh, K., Moghan, S., & Wei, D. L. K. (2007). The situational theory of publics in a different cultural setting: Consumer publics in Singapore. *Journal of Public Relations Research, 19,* 307–332.

Sriramesh, K., & Verčič, D. (2003). *The global public relations handbook: Theory, research, and practice*. Mahwah, NJ: Lawrence Erlbaum.

Triandis, H. C. (1988). Collectivism vs. individualism: A reconceptualization of a basic concept in cross-cultural psychology. In C. Bagley & G. Verma (Eds), *Cross-cultural studies of personality, attitudes, and cognition* (pp. 60–95). London: Macmillan.

Tubilewicz, C. (2008). [Review of the book *China and globalization: The social, economic and political transformation of Chinese society*]. *The China Journal, 59*, 208–210.

Turner, M. M. (2006). Using emotion in risk communication: The anger activism model. *Public Relations Review, 33*, 114–119.

Tyler, L. (2005). Towards a postmodern understanding of crisis communication. *Public Relations Review, 31,* 566–571.

Vardeman, J. E., & Aldoory, L. (2008, Spring). How women make meaning of their shared involvement with spokespersons in news about bioterrorism. *Media Report to Women, 36*(2), 15–22.

Vardeman-Winter, J., & Tindall, N. T. J. (2010). Toward an intersectionality theory of public relations. In R. L. Heath (Ed.), *The Sage Handbook of Public Relations* (pp. 223–235). Thousand Oaks, CA: Sage.

Vercic, D., Grunig, L. A., & Grunig, J. E. (1996). Global and specific principles of public relations: Evidence from Slovenia. In H. M. Culbertson & N. Chen (Eds.), *International public relations: A comparative analysis* (pp. 31–65). Mahwah, NJ: Lawrence Erlbaum.

Wan, H. H., & Pfau, M. (2004). The relative effectiveness of inculation, bolstering, and combined approaches in crisis communication. *Journal of Public Relations Research, 16,* 301–328.

Wang, J. (2008). *Corporate communication practices and trends at the dawn of the new millennium: A China study of 2008.* Retrieved April 8, 2011, from http://www.corporatecomm.org/studies.html

Wang, J., & Goodman, M. B. (2006). *Corporate communication practices and trends: A China benchmark study 2006.* Retrieved April 8, 2011, from http://www.corporatecomm.org/studies.html

Wang, Y., Rees, N., & Andreosso-O'Callaghan, B. (2004). Economic change and political development in China: Findings from a public opinion survey. *Journal of Contemporary China, 13*, 203–222.

Wang, Z. (2005). Before the emergence of critical citizens: Economic development and political trust in China. *International Review of Sociology, 15,* 155–171.

Waymer, D., & Heath, R. L. (2007). Emergent agents: The forgotten publics in crisis communication and issues management research. *Journal of Applied Communication Research, 35*(1), 88–108.

Weber, L., & Parra-Medina, D. (2003). Intersectionality and women's health: Charting a path to eliminating health disparities. In M. T. Segal, V. Demos, & J. J Kronenfeld (Eds.),

Gender perspectives on health and medicine: Key themes, advances in gender research series, 7, 181–230. Oxford, UK: Elsevier Science.

Williams, D. E., & Olaniran, B. A. (1998). Expanding the crisis planning function: Introducing elements of risk communication to crisis communication practice. *Public Relations Review, 24,* 387–400.

Yang, K. S. (1992). Chinese social orientation: From the social interaction perspective. In K. S. Yang & A. B. Yu (Eds.), *Chinese psychology and behavior* (pp. 87–142). Taipei: Laureate.

Chapter 13

Crisis Management and Communication: Pre-Crisis Preparation That Is Sensitive to Diverse Populations

Damion Waymer

The author takes the opportunity in this chapter to extend the concepts of the anticipatory model to include a focus on diversity. He argues that in our changing culture there is a need for public relations experts to be vigilant in pre-crisis planning that recognizes diverse audiences. A compilation of practical considerations for the crisis manager in terms of non-inclusive language and problematic language toward gender, race, and social class is offered here. The author, however, also addresses theoretical considerations for pre-crisis planning in terms of voice and societal relationships.

Introduction

One of the most challenging times for a public relations practitioner to communicate with publics is during a crisis. Organizational control appears to have been lost. Circumstances are uncertain. Situations are often volatile. In short, stakes are high. By definition, a crisis can be considered as a major threat that can have a negative effect on the organization, industry, or stakeholders; that threat, event, and its aftermath can significantly damage the organization experiencing the crisis, its employees, its reputation, financial condition, and products and/or services, if handled improperly (Coombs, 2007). Crisis managers must plan for crises because they are unavoidable. Scholars who study crisis communication think of crises as those predictable yet untimely events that have possible and/or actual consequences for the organization suffering the crisis, the organization's reputation, and its multiple publics, stakeholders, and their interests (Coombs, 1995, 1999, 2007; Seeger, Sellnow, & Ulmer, 1998; Seeger & Ulmer, 2002). Thus, it is not a question of *if* crises will happen, but rather *when* the crisis situations will happen and how to adequately plan for them. This line of logic suggests that although crises themselves are unavoidable, their effects might be minimized if organizations engage in vigilant anticipation of crisis events, scenarios, and their aftermaths.

Vigilant anticipation requires more than just preparing far in advance and polishing a well-crafted speech designed to minimize damage and to address the loss by the organization and/or effected publics immediately following a crisis event (Heath, 2004b). Moreover, anticipation requires more than simply developing a set of strategies designed to repair and restore a tarnished image following a crisis event. Although these two activities are indeed important to crisis management and communication (these activities would occur during the crisis stage and post-crisis stage, respectively), a third-stage set of strategies, activities, and tactics are equally important (these are likely to occur during the pre-crisis stage). Vigilant anticipation in the pre-crisis stage involves calculating risk; conducting risk assessments; determining the potential vulnerabilities of the organization; engaging in an organizational assessment/audit to see which crises are most likely to occur and which crises can be avoided through vigilant preparation; and determining the overall impact of crisis events that are bound to occur.

In recent years there has been a push by crisis communication scholars to pay closer attention to pre-crisis communication and prevention efforts (Coombs, 1999, 2007; Olaniran and Williams, 2001, 2008). Olaniran and Williams (2004) subscribed to the claim that "although one may not be able to prevent all tragedies from occurring, prevention should be a primary concern" (p. 79). Moreover, Olaniran and Williams (2008) asserted that their anticipatory model "views an organization's position within hierarchies as a condition, which has high crisis potential based on the lack of control over factors such as human resources, technologies, and relationship structure" (p. 57). The human resources and relationship structure dimensions provide a ripe opportunity to further apply and advance the reach of their foundational work. This chapter—extending the logic of the anticipatory crisis management model into areas of diversity specifically—establishes that due to the continual influx of ethnically and racially diverse persons in the United States as well as the seemingly academic, cultural, and societal embrace of multiculturalism and pluralism, organizations must be equally (if not more) vigilant in crisis preparation that is sensitive to diverse populations.

Many organizations, especially corporations, likely do not or will not include matters of diversity in their pre-crisis planning (when dealing with internal publics and issues) because it might signal that a racist, sexist, or discriminatory culture exists within the organization. Obviously, few corporate executives view (or wish to view) their organizations in that light. On the other hand, not planning for crises to emerge due to issues of diversity is unwise because many organizations provide services to, employ, and have relationships with publics that are racial minorities (Stewart & Perlow, 2001), members of ethnic minority groups, members that live alternative lifestyles, and/or members who practice various beliefs. Yet, it seems to date that many of these organizations have not found means to strategically manage this diversity in ways that are advantageous for the organization, all the while developing crisis management plans that can prepare for the issues and crises that might arise due to this increased diversity (Waymer, 2010).

Organizations—whether by good deed or by the force of law—have begun to incorporate diversity (and the consequences of not being sensitive to diverse persons' needs) in their operational policies and procedures, but this is still not evidence of pre-crisis planning per se. You see this incorporation of diversity manifested in the increased number of buildings that are accessible for physically disabled persons; it is visible in the increased higher education learning opportunities for those who are learning disabled, visually impaired, and/or hearing impaired; and it is evident in the growing intolerance for issues of sexual harassment and sexual orientation discrimination in workplaces.

This chapter is important because of its proactive stance; it acknowledges that society in general and the workforce specifically are becoming more diverse, and savvy crisis managers and public relations managers would be wise to take into account and plan for this diversity (Waymer, 2010). If organizations are going to deal with internal and external publics that are growing more diverse, it is important that they be able to prepare for and embrace that diversity—or face the potential adverse effects of not doing so.

...Or You'll Be Sorry: Non-inclusive Communication with Diverse Populations

Similar to Seeger's (1986) discussion about how the *Challenger* explosion might have been avoided if certain measures were taken and adhered to, this chapter asserts that due vigilance—in the pre-crisis stage—

must be given to issues and potential crises concerning diverse populations. Diversity should be and must be a focal point of organizations, especially in the pre-crisis stage because diversity—or "difference" as communication scholar Brenda Allen refers to the construct—matters (Allen, 2004). To illustrate this point, government and corporate examples of non-inclusive communication and its effects are provided to lay a foundation and demonstrate the import of sensitive pre-crisis communication with diverse populations. The examples provided will highlight diversity as it relates to gender, race, and class because these three concepts are highly interconnected and have been considered the most enduring and divisive social constructions (Aldoory, 2009; Cloud, 1996; Giroux, 2003).

Matters of Problematic Gendered Communication

Allen (2004) argued that humans are not born into a gender; "rather we learn gender through interactions with others and are socialized to assume a gender role, which we continue to perform throughout our lives" (p. 40). Since we enter a world that is coded and structured in terms of gender from birth, some public relations scholars argue that gender is likely "the most restricting force in American life" (Aldoory, 2009, p. 110).

One noted example of the restrictive power of gender occurred in 1996 when actress Hunter Tylo was fired from Fox Network's *Melrose Place* after she told producers that she was pregnant. Hunter Tylo, who did not appear in an episode at the time, sought at least $2.5 million in damages for her claims of pregnancy discrimination and wrongful termination and breach of contract (Jennings, 2009). Producers argued during the trial that Tylo violated the terms of her contract by getting pregnant because her contract prohibited her from changes in her appearance, and pregnancy would indeed result in a material change; moreover, they argued that Tylo's pregnancy would have made her too large and not physically suited to play the role of the sexy conniving seductive woman (Jennings, 2009).

In rendering its verdict, the jury found that the show's producers' arguments—that Tylo's firing was justified on the basis that, while pregnant, she could not convincingly play her role as a vixen on the nighttime soap opera—to be unacceptable. As a result, a jury awarded Hunter Tylo around $4.8 million for emotional distress and economic loss she suffered after being fired from the Fox Network TV show (Jennings, 2009).

Although neither Fox nor *Melrose Place* producers were crippled by this lawsuit, it can still be deemed a crisis. If this were a smaller company, this lawsuit might have led to the company's demise. Nearly $5 million is not a small amount of money to lose, and more importantly a tarnished reputation takes time to repair. Simply put, male actors are not likely to be subject to gender discrimination based on pregnancy status. Thus, this situation could have been handled in vastly different ways that would show the production company's and the industry's sensitivity towards diverse populations.

Another example is the more recent incident with Pepsi, AMP Energy drink, and their joint iPhone application called "Before You Score"—with "score" meaning entering a sexual relationship with a woman. The application, which categorized women into 24 types, suggested the right words to use to "score" with each type of woman. If a person is lucky enough to "score" with a woman, the application encourages the person to boast of the conquest via e-mail, Facebook, or Twitter. AMP has always targeted men in its communications; however, women who came across this application took serious issue with AMP, Pepsi, and the application. Pepsi and AMP quickly issued an apology for the offensiveness of the application, finally pulling it after two weeks. In this example, Pepsi and AMP attempted to communicate with a specific stakeholder group (young men) without considering the implications of issuing a message that is insensitive, inconsiderate to other marginalized publics, and inconsistent with what society deems to be acceptable behavior (Boyd & Waymer, 2011).

In both examples, persons either did not see or chose to ignore the potential crises on the horizon. They did not see or miscalculated the likelihood and potential impact of their organizational decisions—which resulted in a crisis situation. A careful articulation of diversity in Pepsi's and AMP's pre-crisis planning would have accounted for the potential public outrage behind what appears to be a gendered discriminatory, sexist application for a smart phone device. Likewise, Spelling Productions (producers of *Melrose Place*) could have warded off potential bad press, a lawsuit, and loss of millions had they considered that the young women they work with could eventually become pregnant. The problem, however, was that these producers were blinded by a prevailing dominant ideology: the ideal worker ideology (Williams, 2000).

The ideal worker is someone who can and will put his or her career and the rewards that it brings first. According to this definition, ideal workers can be either male or female—though, due to gender roles men are still widely regarded as ideal workers—as long as they do not have to devote large amounts of their time to other duties (i.e., familial obligations such as childcare and care for ailing parents and other family members) (Williams, 2000). In short, pregnancy is not a "bona fide occupational qualification" (Jennings, 2009)—meaning employment cannot and should not be predicated on pregnancy status. As such, any policy or action that appears to terminate a person's employment based on her/his pregnancy status is grounds for an issue at minimum, a crisis at worst, to arise.

Matters of Problematic Racial Communication

Waymer (2010) argued that race merits theoretical and practical study because it is an enduring, contested phenomenon with important communicative implications. Moreover, he asserted that scholars of crisis communication and public relations must move beyond seeing race as variable—something to be instrumentalized, something to be managed by, or something to be managed through public relations—and get to a point where we see public relations as being a theoretically rich vehicle to address an enduring construct that can advance or impede the vision community that Heath (2006) called for in his Fully Functional Society Theory (FST).

In November of 2000, Coca-Cola agreed to pay the largest sum ever paid in a racial discrimination case, $192.5 million, to current and former employees who were plaintiffs in *Ingram et al. vs. The Coca-Cola Company* (Parker, 2006). This case originated in 1995, when Coca-Cola made Carl Ware its senior vice president. Ware teamed with other African American employees to document what they felt were shortcomings with Coca-Cola's diversity plan. According to what became known as the "Ware Report," Coca-Cola's diversity plan failed in many areas, such as including diversity at the decision-making level, respecting different cultures, and giving minority employees opportunities for advancement (Parker, 2006). According to the testimony from *Ingram et al. vs. The Coca-Cola Company*, Coca-Cola's board ignored the report and did not address any of the concerns presented by Ware and the other employees. This led to a group of then-current and former Coca-Cola employees filing racial discrimination charges against Coca-Cola in April of 1999 (Parker, 2006).

While Coca-Cola was initially able to gain support from leaders in the minority community because of its history of contributing to and working with minority organizations, knowledge that Coca-Cola had been aware of the Ware Report for years damaged Coca-Cola's reputation. In addition to damaging its image, it can be speculated that Coca-Cola also suffered financially because of the case. Between 1998 and 2000 Coca-Cola's stock price dropped $100 billion. A major consulting firm determined that Coca-Cola's brand rating dropped from 82.7 in 1995 to 77.1 in 1999. Coca-Cola settled the case in 2000 (Parker, 2006).

This case is highlighted, however, due to its import and implications for post-crisis image management and the potential for pre-crisis management. After this groundbreaking case, Coca-Cola used its website to:

1. showcase its efforts to change its corporate culture after the lawsuit and its commitment to the minority community through financial donations, minority hiring, and other programs.
2. embrace diversity as a core value.
3. communicate corrective action; Coca-Cola showed changes it made to its company on its website, including surveys that documented the effectiveness of its diversity plan and Task Force Reports that showed how well it was implementing the terms of the lawsuit.
4. create distance (from the problem) through bolstering; Coca-Cola's diversity website positively portrays its relationship with the minority community and shows the strides the company has made since the lawsuit (Parker, 2006).

Although Coca-Cola has made efforts to address its racial crisis, the company still has suffered image damage due to racial issues. In early 2010, Sprite (a trademarked brand of the Coca-Cola Company) hosted a national step show—the largest Greek step show ever. This step show received national exposure, and MTV2 televised the event. Stepping, a demonstrative art of dance and military type cadences that originated in Black Greek Letter Organizations (BGLOs), is the trademark of black sororities and fraternities. The irony of this crisis, however, is that Coca-Cola via Sprite appeared to be making a genuine effort to foster mutually beneficial relations with African American communities. Sprite was offering scholarship money to Historically Black Colleges and Universities (HBCUs) via this step show; Sprite, with the help of MTV, was giving this unique art form a national platform. The problem that arose in this case is that on February 20, 2010, at the national finals of the Sprite Step Off competition, the all-white college girl team from Zeta Tau Alpha of the University of Arkansas at Fayetteville won the sorority division championship and took home the first-place prize of $100,000. Black sorority sisters and many people in the black community were upset (Tucker, 2010). Part of the controversy arose because MTV2 followed the sisters of Zeta Tau Alpha with camera crews, and many felt it was a conspiracy that they won the competition. Others were simply upset that white girls were competing in what is considered "black art":

> When it was announced that the Zetas won, the feel-good vibe evaporated. Large sections of the crowd starting booing. Then Internet and radio-call-in warfare broke out when the videos were posted on YouTube. There were allegations of cultural theft and reverse racism, not to mention race-based taunting and name-calling. (Tucker, 2010)

Just reading some of the comments posted on the YouTube site, one can witness an array of comments about the events from this step show. As of September 2010, the video has received more than 821,000 hits, and persons are still posting their comments and adding to the discussion.

Some might ask why this is an issue for Coca-Cola and Sprite. About one week after the step show finale (and the scourge of outcry), Sprite officials announced that they discovered a scoring discrepancy:

> This was odd because the show's host, rapper Ludacris, assured the crowd that the judges' scores had been "double-checked."...Spokesmen from the Coca-Cola Co. (Sprite's corporate parent) won't say anything beyond a vague official statement: "We conducted a post-competition review and discovered a scoring discrepancy. There is no conclusive interpretation, nor definitive resolution for the discrepancy." Thus, Coca-Cola gave both groups $100,000. (The second-place prize had been $50,000.) It's also fair to note that contest officials are so tight-lipped about the event that they would not identify the judges they hired for the competition, even though it was filmed for television in front of more than 4,000 people. (Tucker, 2010)

Coca-Cola decided to simply pay an additional $50,000 in an attempt to make this crisis go away. Although this can be deemed as a form of corrective action, it might not have been the best course of action. This act appeased few publics. Maybe there was a scoring discrepancy. However, if there wasn't, Coca-Cola missed a golden opportunity to put its commitment to diversity into action. Instead of writing a check to silence critics, Coca-Cola could have used this platform to engage a critical discussion about race and relations. Other companies can learn a lesson from Coca-Cola's mishandling of this issue. Having a more nuanced understanding of race and creating pre-crisis plans and actions rooted in that understanding are indeed a lesson for crisis communication managers and applied communication researchers.

Matters of Problematic Social Class Communication

Social class "embodies a powerful, persistent predictor of accessibility to resources, potential for longevity and success, and self-esteem" (Allen, 2004, p. 99). Class is not just about economic status but social status as well:

"Believing there is more to class than just economics, Weber maintained stratification is based on property, power, and prestige (the three Ps)." (Allen, 2004, p. 98). The American Dream ideological myth of success is the embodiment of the three Ps, and although few Americans' lives actually exemplify the American Dream, all forms of media perpetuate the myth that achieving the American Dream is within everyone's grasp. Gray (1989), in his essay about television, black Americans, and the American ideology of success, stated that "one message of these representations of success and failure is that middle…[and upper class] class blacks (and members of all other races and ethnic groups) succeed because they take advantage of available opportunities while poor blacks and other marginal members of our society fail because they do not" (p. 376). This is problematic because this can lead to a belief that people overwhelmingly are to blame for their "non-success"—without taking into account the socially constructed barriers present within a given society. By refusing to acknowledge, account for, and address, in part, the political, economic, and structural barriers that physically and psychology separate an organization from its myriad publics, an organization sustains a suitable environment for a simmering, smoldering crisis to brew and eventually erupt.

For example, On April 7, 2001, Timothy Thomas, an unarmed black male, was shot by a white Cincinnati police officer. Thomas was the fifteenth African American male to be killed by Cincinnati police since 1995. Within days, black Cincinnatians were protesting and rioting. Many believed that the city's riots were a manifestation of Cincinnati's long history of police brutality toward African Americans, but the reality is that the police had been playing the role of protector of property—keeping the material possessions that the "haves" have from the "have-nots"—which is indeed a social class contestation (White, 2001). And governments will be called upon to decide how to best manage these conflicts and balance competing interests (Waymer, 2009).

More than four decades ago, on June 12, 1967, black Cincinnatians, displeased with the circumstances in which they lived, protested and rioted; less than a year later, in April 1968, Cincinnati was again torn by riots following the assassination of civil rights leader Martin Luther King, Jr. In response to the riots, the Johnson administration's Commission on Civil Disorders issued a report, the Kerner Report, and concluded that the following events greatly contributed to Cincinnati's riots: (1) the impoverishment of Cincinnati's segregated neighborhoods; (2) police officers' practice of stopping African Americans without obvious grounds; and (3) police officers' use of loitering laws disproportionately against people of color. Although rioting ceased at the turn of the decade, the root causes—the three Ps of lacking property, power, and prestige—that had contributed to the rioting remained unaddressed (White, 2001). Those issues remained relatively unaddressed in 2001 as well.

This brief account of the crisis that Cincinnati faced and the conditions leading up to the riot of 2001 demonstrate what happens in most crisis situations to date: "the conditions leading to a crisis linger, smolder for a time, until something causes them to flare" (Millar, 2004, p. 27). In this particular situation the flare was the murder of Timothy Thomas, who just so happened to be the fifteenth African American male killed by the police in a six-year period.

This riot was a public relations nightmare for the city. The city had to recognize that it had serious challenges following this crisis. To respond to these challenges, the city welcomed the presidential-appointed task force, and then the city's spokesperson—Mayor Charlie Luken—vowed to clean up the blighted inner-city Over-the-Rhine area (WCPO, 2002). And in 2003, Cincinnati government, along with local business leaders, hired John Alschuler—a New York consultant—to devise a development plan for Greater Cincinnati's core (Horstman, 2003). These actions alone, however, were not enough. Cincinnati pulled together as a city to address those smoldering issues. Crises management is an ever-going process.

Coombs (1999) stated a crisis must be monitored when it is over, "even if no changes are created" (p. 146). He further asserted:

> Crisis tracking monitors factors that produced the crisis to see if another threat may arise. Crisis tracking feeds back into signal detection and crisis preparation. Simply put, the postcrisis phase ends with the crisis managers moving back to the actions involved in the precrisis phase of crisis management. (p. 146)

Crises create rhetorical problems that demand that sense be made of decisions and events. Moreover, "the organization suffering the crisis is expected to understand and respond to the various kinds of sense that are being made" (Heath, 2004a, p. 249). In a country in which its history is filled with socioeconomic disparity, organizations must be vigilant about identifying and planning for crises that might emerge due to social class conflicts—especially in down global economic times.

Simply put, matters of diversity are important. They must be given greater attention in pre-crisis communication planning, management, and communication. It is better to be proactive and have a plan in place (and be considered cynical for believing that discriminatory practices might befall your organization one day) than to not have a plan in place when suddenly a crisis of diversity emerges and damages your organization via bad press, lawsuits, or even a riot. Ignoring the elephant (diversity) in the room does not make the elephant magically disappear. It just ensures that you and others might get trampled if and when the elephant decides to charge (a crisis of diversity emerges). Talk about the elephant. Try to embrace the elephant. Make a plan for the elephant.

A Development of Pre-Crisis Preparation That Is Sensitive to Diverse Populations

Millar (2004) found that different industries suffer different crises—meaning that certain organizations, depending on their operating procedures and the nature of business in which they are involved, are more prone to suffer certain crises. For example, he found that health care organizations tend to experience high amounts of white-collar crime and mismanagement, whereas "sexual harassment becomes more frequent in higher education than either among organizations [studied] or health care" (Millar, 2004, p. 27). This leads to the position that organizational culture, which can be influenced by industry affiliations, can be a fruitful avenue to extend pre-crisis communication research by prescribing industry specific pre-crisis strategies to deal with crises involving matters of diversity. Hence, organizations should not ignore the research; they should accept the fact that certain industries are more susceptible to crises involving matters of diversity and make pre-crisis plans accordingly.

After issues with the potential of becoming crises, such as issues of diversity, have been identified, scanned, monitored, and analyzed, crisis managers should apply a systematic approach to plan for the issue, to prepare better for the issue, and to resolve, potentially, the issue in a manner favorable to the organization (Heath & Palenchar, 2009). The Jones and Chase (1979) model, a dominant issues management model, provides "crisis managers with directions on actions to take to prevent an issue from becoming a crisis and ways to evaluate the issues management effort" (Coombs, 1999, pp. 40–41). Hence, risk assessment as well as strategic issues management can be thought of as an indispensable pre-crisis procedure (Coombs, 1999, 2007). In short, "vigilant preparation (by self-reflective organizations) can reduce the likelihood of a (risk manifesting itself into a) crisis and increase the responsiveness of the organization to demonstrate its ability to establish control over its operations" (Heath, 2004b, p. 35).

Additionally, organizations should analyze and, if needed, revise their culture and alter corporate governance if necessary. In the aforementioned examples, there was (and some would say continues to be) a culture of unwritten gender discriminatory procedures in the film and entertainment industry that locks actresses into roles. There was rampant racism at Coca-Cola, and the organization chose to ignore the Ware Report. There was a culture of class-based discrimination and economic disparity in the city of Cincinnati for many years (Waymer, 2009). Diversity, in short, oftentimes is a matter of culture. Diversity has to be a part of the organizational culture if it is to be embraced by organizational members.

Culture is an important part of any organization. Although definitions vary, it is generally agreed that culture spreads through an organization via stories and other artifacts (Smith & Keyton, 2001). Culture has a widespread and profound effect on the attitudes and beliefs of organizational members. As stories filter through, people gain a sense of acceptable and unacceptable behaviors, which helps define the power and politics in operation; in an organizational context, the interpretation of these messages can affect employees' sense of empowerment by helping or hindering a sense of belonging and unity (Fairhurst & Putnam, 2004; Smith & Keyton, 2001; Taylor, Irvin, & Wieland, 2006).

Crises and culture can be interrelated, and because a corporation's culture permeates an organization's daily operations, culture has been studied and found to be a contributing factor to organizational crisis (Bechler, 1995, 2004) and the after-effects of crisis on image and reputation (Lyon & Cameron, 2004). Simply put, scholars have shown that culture can lead to crisis. Thus, it is unwise not to prepare for crises of culture—and issues of diversity should be at the fore.

Theoretical and Applied Implications

Crisis communication can be viewed as a highly specialized blend and form of public relations, organizational, and managerial communication. As such, crisis communication is subject to many (if not all) of the problematics present in this arena of study. Mumby and Stohl (1996) argued that there are four major problematics in organizational communication, and most (if not all) of the research conducted in this area falls under one of these problematics. They are the problematic of voice, the problematic of rationality, the problematic of organization, and the problematic of the organization–society relationship. Three of these four problematics can be applied to crisis communication and diversity to help us understand better the possibilities of developing pre-crisis plans and communications that are sensitive to diverse populations.

The first relevant problematic—the problematic of voice—deals with the various voices and standpoints that members of organizations represent. Some studies address the managerial voice while others focus on marginalized voices, silenced voices, and/or competing voices. Due to the increased complexity of organizations and the diversity of their members, it is likely that there exist multiple accounts of what occurs within organizations, how policies are viewed, and what is acceptable. Diversity, in this vein, is the manifestation of these voices. Multiple voices add value to dialogue. In fact, scholars have argued that diversity in public relations adds value to society's discourse because it increases the likelihood that concerns will be heard and given regard (Hon & Brunner, 2000). From a rhetorical standpoint, is there anything more ethical than saying, *let the dialogue continue?* (Heath, 2009). And as Waymer (2010) noted, that dialogue on diversity must continue, and it must be meaningful.

The next relevant problematic—the problematic of rationality—according to Mumby and Stohl (1996), can be viewed in two parts: the technical rationality is the rationality that deals with predicting, explaining, and controlling phenomena studied within organizations while the other rationality deals with studying members of organizations and seeing how they make sense of their surroundings and individual experiences. Both rationalities are important to organizational communication scholarship in general and pre-crisis communication specifically, although scholars are, at times, at odds on these issues of rationality. The entire premise of pre-crisis communication is to use social scientific methods to predict the likelihood that a crisis will occur, as well as the impact that crisis might have on publics and the organization that would suffer from the crisis. Crises are inevitable, yet scholars attempt to control the elements and multiple variables that can make a bad crisis situation a worse situation. It makes good business sense, and it is socially responsible. The second part of the problematic of rationality also can play a key role in pre-crisis management—especially as it pertains to issues and crises of diversity. It is important to know how diverse persons make sense of situations, how they prefer to be communicated with, under what conditions it is best to communicate with them, etc.

Coombs (2009) suggested that it "may be time to combine the rhetorical approach with social scientific approaches, as in SCCT [Situational Crisis Communication Theory], to broaden our field of view and to systematically test our recommendations for crisis response and post-crisis communication" (p. 249). In essence, he is arguing for at least a two-prong application to tackle crisis communication that draws upon the strength of both components of the problematic of rationality. Not only can his recommendations be used to strengthen pre-crisis planning, but by being "diverse" in methodological approaches, one might be able to better plan for, account for, and embrace issues and crises of diversity before they arise. For example, Heath, Lee, & Ni (2009) surveyed more than 600 white, black, Asian American, and Hispanic Americans in the greater Houston, Texas, area. What they found was that about half of the people surveyed did not feel adequately prepared to respond during times of crisis such as a chemical plant explosion, a natural disaster, or terrorist attacks. And for those who did feel adequately prepared to respond, members of minority publics voiced that diverse voices (source similarity and message sensitivity) increased residents' sense of self, expert, and community preparedness. In short, if citizens have access to risk and crisis information that is from sources similar to them and stated in messages that are sensitive to them, they feel more prepared to deal with crisis emergency response. This line of research, focused on being more inclusive to welcome perspectives from minority perspectives so that there can be further efforts to develop better pre-crisis plans and communication that would be sensitive, inclusive, and responsive to the needs of diverse populations, needs to be conducted in greater numbers so that scholars and practitioners alike can have a firm base to implement crisis response messages for diverse populations that are rooted in sound theory and good research.

The third and final problematic, the problematic of the organizational-society relationship presented by Mumby and Stohl (1996), addressed the interconnectedness of organizations and the societies in which they operate. Within this problematic, scholars are demonstrating how organizations influence their societies and how societal constraints shape and influence these organizations. The line between these two—organizations and society—is blurred; often one bleeds into the other and vice-versa. Nowhere else is this problematic more evident than in studying public relations and crisis communication. When organizations suffer a crisis—the Toyota sticking accelerator recall, Mattel lead paint toy recall, the U.S. government and its response to Hurricane Katrina, the BP oil spill—myriad stakeholders oftentimes also suffer from that crisis, whether it be inconvenience, their livelihoods, or even their lives.

Conclusion

Organizations must take into account—before a crisis occurs—the effects of their actions. And due to the increased diversity present in the U.S., as well as the increased connectivity of the global world, ignoring matters of diversity simply is not an option. Management must be reflective. Management, in its planning, operations, policies, and procedures, must consider the "the voices of those who might be forgotten but whose interests are affected by the organization. Being reflective, inclusive, and proactive can enhance their crisis response and reduce indifference to either the marginalized publics or their suffering" (Waymer & Heath, 2007, p. 106). Can management consider itself socially responsible or ethical if they are not engaging constantly in reflective practices that consider the perspectives of the diverse populations?

References

Aldoory, L. (2009). Feminist criticism in public relations: How gender can impact public relations texts and contexts. In R. L. Heath, E. L. Toth, & D. Waymer (Eds.), *Rhetorical and critical approaches to public relations II* (pp. 110–123). New York: Routledge.

Allen, B. J. (2004). *Difference matters: Communicating social identity*. Long Grove, IL: Waveland.

Bechler, C. (1995). Looking beyond the immediate crisis response: Analyzing the organizational culture to understand the crisis. *Journal of the Association for Communication Administration, 1,* 1–17.

———. (2004). Reframing the organizational exigency: Taking a new approach in crisis research. In D. P. Millar & R. L. Heath (Eds.), *Responding to crisis: A rhetorical approach to crisis communication* (pp. 63–74). Mahwah, NJ: Lawrence Erlbaum.

Boyd, J., & Waymer, D. (2011). Organizational rhetoric: A subject of interest(s). *Management Communication Quarterly, 25*, 474–493.

Cloud, D. L. (1996). Hegemony or concordance? The rhetoric of tokenism in "Oprah" Winfrey's rags-to-riches biography. *Critical Studies in Mass Communication, 13,* 115–137.

Coombs, W. T. (1995). Choosing the right words: The development of guidelines for the selection of the "appropriate" crisis-response strategies. *Management Communication Quarterly, 8,* 447–476.

———. (1999a). *Ongoing crisis communication: Planning, managing, and responding.* Thousand Oaks, CA: Sage.

———. (1999b). *Ongoing crisis communication: Planning, managing, and responding*. (2nd ed.). Thousand Oaks, CA: Sage.

———. (2009). Crisis, crisis communication, reputation, and rhetoric. In R. L. Heath, E. L. Toth, & D. Waymer (Eds.), *Rhetorical and critical approaches to public relations II* (pp. 237–252). New York: Routledge.

Fairhurst, G., & Putnam, L. (2004). Organizations as discursive constructions. *Communication Theory, 14,* 5–26.

Giroux, H. A. (2003). Spectacles of race and pedagogies of denial: Anti-Black racist pedagogy under the reign of neoliberalism. *Communication Education, 52,* 191–211.

Gray, H. (1989). Television, black Americans, and the American Dream. *Critical Studies in Mass Communication, 6,* 376–386.

Heath, R. L. (2004a). After the dance is over: Postcrisis response. In D. P. Millar & R. L. Heath (Eds.), *Responding to crisis: A rhetorical approach to crisis communication* (pp. 247-249). Mahwah, NJ: Lawrence Erlbaum.

———. (2004b). Crisis preparation: Planning for the inevitable. In D. P. Millar & R. L. Heath (Eds.), *Responding to crisis: A rhetorical approach to crisis communication* (pp. 33–35). Mahwah, NJ: Lawrence Erlbaum.

———. (2006). Onward into more fog: Thoughts on public relations' research directions. *Journal of Public Relations Research, 18*(2), 93–114.

———. (2009). The rhetorical tradition: The wrangle in the marketplace. In R. L. Heath, E. L. Toth, & D. Waymer (Eds.), *Rhetorical and critical approaches to public relations II* (pp. 17–47). New York: Routledge.

Heath, R. L., Lee, J., & Ni, L. (2009). Crisis and risk approaches to emergency management planning and communication: The role of similarity and sensitivity. *Journal of Public Relations Research, 21,* 123–141.

Heath, R. L., & Palenchar, M. J. (2009). *Strategic issues management: Organizations and public policy challenges* (2nd ed.). Thousand Oaks, CA: Sage.

Hon, L. C., & Brunner, B. (2000). Diversity issues and public relations. *Journal of Public Relations Research, 12,* 309–340.

Horstman, B. M. (2003, June 5). Expert: Arts drive city's future. *The Cincinnati Post,* p. A1.

Jennings, M. M. (2009). *Business ethics: Case studies and selected readings* (6th ed.). Mason, OH: Cengage Learning, Inc.

Jones, B. L., & Chase, W. H. (1979). Managing public policy issues. *Public Relations Review, 5,* 3–23.

Lyon, L., & Cameron, G. (2004). A relational approach examining the interplay of prior reputation and immediate response to a crisis. *Journal of Public Relations Research, 16,* 213–241.

Millar, D. P. (2004). Exposing the errors: An examination of the nature of organizational crises. In D. P. Millar & R. L. Heath (Eds.), *Responding to crisis: A rhetorical approach to crisis communication* (pp. 19–31). Mahwah, NJ: Lawrence Erlbaum.

Mumby, D., & Stohl, C. (1996). Disciplining organizational communication studies. *Management Communication Quarterly, 10,* 465–486.

Olaniran, B. A., & Williams, D. E. (2001). Anticipatory model crisis management. A vigilant response to technological crises. In R. Heath (Ed.), *Handbook of public relations* (pp. 487–500).

———. (2004). Burkian counternature and the vigilant response: An anticipatory model of crisis management and technology. In D. P. Millar & R. L. Heath (Eds.), *Responding to crisis: A rhetorical approach to crisis communication* (pp. 75–94). Mahwah, NJ: Lawrence Erlbaum.

———. (2008). Applying anticipatory and relational perspectives to the Nigerian delta region oil crisis. *Public Relations Review,* 34, 57–59.

Parker, P. S. (2006). Keeping it real: Race, difference, and corporate ethics at Coca-Cola. In S. May (Ed.), *Case studies in organizational communication: Ethical perspectives and practices* (pp. 169–184). Thousand Oaks, CA: Sage.

Seeger, M. W. (1986). The Challenger tragedy and search for legitimacy. *Central States Speech Journal, 37,* 147–157.

Seeger, M. W., Sellnow, T. L., & Ulmer, R. R. (1998). Communication, organization, and crisis. In M. E. Roloff (Ed.), *Communication yearbook 21* (pp. 231–275). Thousand Oaks, CA: Sage.

Seeger, M. W., & Ulmer, R. R. (2002). A post-crisis discourse of renewal: The case of Malden Mills and Cole Hardwoods. *Journal of Applied Communication Research, 30,* 126–142.

Smith F., & Keyton, J. (2001). Organizational storytelling: Metaphors for relational power and identity struggles. *Management Communication Quarterly, 15,* 149–182.

Stewart, L. D., & Perlow, R. (2001). Applicant race, job status, and racial attitude as predictors of employment discrimination. *Journal of Business and Psychology, 16,* 259–279.

Taylor, B. C., Irvin, L. R., & Wieland, S. M. (2006). Checking the map: Critiquing Joanne Martin's metatheory of organizational culture and its use in communication research. *Communication Theory, 16,* 304–332.

Tucker, N. (2010, March 4). Sprite Step Off: In step with the times, or going a step too far? *The Washington Post.* Retrieved from http://www.washingtonpost.com/wp-dyn/content/article/2010/03/03/AR2010030303895.html

Waymer, D. (2009). Liberty and justice for all? The paradox of governmental rhetoric. *Communication Quarterly, 57,* 334–351.

———. (2010). Does public relations scholarship have a place in race? In R. L. Heath (Ed.), *The Sage handbook of public relations* (pp. 237–246). Thousand Oaks, CA: Sage.

Waymer, D., & Heath, R. L., (2007). Emergent agents: The forgotten publics in crisis communication and issues management research: *Journal of Applied Communication Research, 35*, 88–108

WCPO (2002). Mayor reacts to Vine 1/10/02 [Electronic Version]. *WCPO—TV in Cincinnati Local News and Segments.* Retrieved from http://web.archive.org/web/20020619133922/http://www.wcpo.com/wcpo/localshows/iteam/visionsofvine/4c639b.htmlhttp://www.theolympian.com/2010/06/16/1275880/bp-agrees-to-set-aside-20-billion.html

White, J. (2001). The Cincinnati riots and the class divide in America: Gentrification and police repression. The World Socialist Web Site. Retrieved from http://www.wsws.org/articles/2001/may2001/cin1-m24.shtml

Williams, J. (2000). *Unbending gender: Why family and work conflict and what to do about it.* Oxford, NY: University Press.

Chapter 14

Racially Oriented Crises: A Unique Challenge to Crisis Preparation

David E. Williams & Bolanle A. Olaniran

In this chapter, the authors coin the term ROCs (racially oriented crises). ROCs are presented as crisis types that have unique characteristics, which require unique anticipatory preparations. They argue that ROCs are unique because they are difficult to anticipate, result in reputational damage, and tend to have responses with polarized strategy options. The chapter offers specific insights into how organizations can best prepare for potential ROCs.

Introduction

Racially oriented crises (ROCs) are among the most difficult to manage and potentially devastating of all crisis types. Ulmer, Sellnow, and Seeger (2007) identify crises as being intentional (e.g., terrorism or sabotage) or unintentional (e.g., natural disasters or downturns in the economy). Racially oriented crises of the past have proven to be either or both types. Crises involving race can range from as a single spoken ethnic slur to a policy or practice that invokes depictions of a history of bias and prejudice.

Baker (2001) helped operationally define ROCs when she identified "charges of racism against an organization, use of racial slurs by corporate officials or other employees, lawsuits alleging discrimination on the basis of race" (p. 514). From Baker's parameters, it is clear that ROCs can stem from any level of an organization and can result from words, actions, or both.

Baker also identified prominent organizations that have been involved with ROCs. She noted Texaco, American Airlines, Pillsbury, AT&T, General Motors, Eddie Bauer, and Denny's. To that list, we can add names of individuals such as Don Imus, Michael Richards, and Rush Limbaugh. Entities dealing with racial crises also include politicians and municipalities at all levels dealing with incidents of police actions, voting allegations, and decision-making challenged on premises involving race or discrimination. One quickly develops an accurate perception of ROCs as being all-encompassing, ever-present, and without boundaries in American business and culture.

Uniqueness of ROCs

Three qualities make ROCs unique and difficult to manage: These crises are difficult to anticipate, the damage targets the organization's or individual's reputation, and the responses tend to require the crisis response team to choose among polarized strategies.

Racially oriented crises can emanate from behaviors that are culturally inappropriate or insensitive. These race-based crises are not as difficult to anticipate, yet they still occur and leave organizations struggling to respond. Denny's restaurant stands as an example with cases in the early 1990s of discrimination based on race by denying service to minority group customers. These incidents resulted in lawsuits including a $54 million settlement in 1994 (Adamson, 1998; Chin et al. 1998). While the refusal of service is not difficult to recognize as a potential crisis trigger, it still happened. The distance between Denny's management and individual restaurant chains allowed for a string of high-profile incidents to erupt before the organization could put the crisis behind them.

Frequently, ROCs stem from language choices made by someone within an organization. Whether we are speaking of a celebrity with constant media attention or high-level management members of an organization, the spoken word can haunt the speaker or organization while igniting a crisis.

Texaco housed such a crisis in 1996 when Richard A. Lundwall released a tape of a conversation with three other executives (Brinson & Benoit, 1999). The tape revealed comments referring to African Americans as "black jelly beans" being "glued to the bottom of the jar" among other statements that reflected poorly on the executives who uttered them. While the language is inflammatory, the anticipation of this crisis was difficult. Texaco's crisis preparation likely focuses on the possibility of product failure or industry accidents. The need to monitor the talk of high-level executives was not likely a high priority. The San Francisco 49ers professional football team also found itself in the midst of a crisis, which touched on a few hot button issues, including race and an inappropriate Asian stereotype. As the training video at the heart of the crisis was actually created by a member of the 49ers public relations staff, it stands as further evidence that ROCs are difficult to anticipate.

With Texaco and the 49ers crises, it would seem common sense should have prevailed and these crises would have been avoided. However, even public relations personnel sometimes fail to monitor their communication. When the failure results in a racially offensive comment, the potential for crisis is present. If high-level executives at a major company and a public relations professional can produce crisis-instigating communication, so can many other members of organizations. It is virtually impossible to monitor all communication from anyone who could later be identified as the dreaded "spokesperson for the organization."

Racially oriented crises are also unique in terms of how organizational or individual reputation usually pays the greatest price for the action or language. Crises that focus on race can entice a negative response from a wide range of stakeholders. Media, groups representing the targeted groups, and large segments of the general public will all typically respond to racial incidents. While all crises affect the reputation of an organization, ROCs cut to the moral character of the organization and its leadership. In both the Denny's and Texaco cases, responses from media and angry stakeholders challenged how the leadership could allow such a culture to exist in their companies. The assumption is made that those at the top of the hierarchy reflect, or at least refuse to condemn, the attitude that allows racial comments or acts to ensue.

The damage to reputation is most evident in celebrity cases of racially oriented crises. Don Imus (Carr, 2007) and Michael Richards (Richards hires) both experienced significant damage to their professional and personal reputations from comments made regarding African Americans. In both cases, their comments resulted in immediate characterizations of them as racists. There was little attempt by the media or responding stakeholders to separate the person from the words and suggest that the incidents were exceptions to their character. In the case of Imus, a common reaction was to report not only the incident but also previous incidents, which would seem to correlate the assessment of his comments regarding the Rutgers women's basketball team as evidence of his racist attitudes (e.g., Racism is to be).

Organizations also feel the reputational sting of ROCs as evidenced by two restaurant chains. Denny's was not alone in managing the challenges faced by allegations of racial discrimination. In 1989 Shoney's was attacked by the NAACP for discriminatory practices. The organization was charged with limiting employment opportunities and advancement for minorities and creating a hostile work environment. Shoney's eventually settled the case for well over $100 million. However, the stigma of the racial charges followed the company well into the 90s, despite their efforts to recruit, retain, and advance minority applicants and support significant community involvement.

Attacks on individual or organizational character, stemming from ROCs, tend to generate more negative responses from the general public and stakeholders. While protests and strikes are typical outlets for groups angered by any type of crisis, ROCs join with environmental crises in generating a more significant response from a wide range of affected or angered stakeholders and segments of the public.

The need to more frequently respond with an extreme strategy constitutes the third unique characteristic of ROCs. Crisis response strategies can range from those that largely acquiesce to the allegations involved in the crisis to those that forcefully challenge the accuracy, motive, or existence of the crisis. Coombs (2007) offers a useful typology with the Situational Crisis Communication Theory (SCCT). The SCCT included three postures of denial: diminishment, rebuilding, and bolstering. Within the postures are ten strategies: attacking the accuser, denial, scapegoating (denial), excusing, justification (diminishment), compensation, apology (rebuilding), reminding, ingratiation, and victimage (bolstering).

Individuals and organizations facing ROCs tend to respond with one of the more extreme strategies identified in this typology. The rebuilding posture strategies of compensation and apology are frequently used in response to ROCs. Baker (2001) suggests:

> In situations involving race, it might be prudent for public relations professionals to advise management to immediately acknowledge offensive behavior, take swift correction action, and review all company policies in an effort to prevent such incidents from arising again. (p. 520)

Baker addressed one ROC case (Eddie Bauer) in which decision-makers did not follow the trend to respond in either extreme and opted for denial and evasion. In this case, an African American teen was in an Eddie Bauer store wearing an Eddie Bauer shirt purchased the previous day. He was approached by a security guard who asked him to produce a receipt. Having no receipt, the teen was told to remove the shirt and go home to get the receipt.

The company responded to charges of racism by suggesting it was a minor issue and it was a police matter. These responses were not viewed positively by the general public or specific groups. Attempts to downplay or step away from ROCs will typically fail. Therefore, responses that follow the suggestions of Baker are more typical. Texaco, in responding to the taped executives' comments about African Americans, relied on a range of strategies, more in line with Baker's recommendations. In Brinson and Benoit's (1999) terms, Texaco used bolstering, corrective action, mortification, blame shifting, and separation. Corrective action and mortification (e.g., accepting the blame and apology) represent crisis response decision-making that admits guilt and attempts to prevent the return of the problem. The authors' note of Texaco's use of bolstering, correction action, and mortification "[t]ogether, these image repair strategies, along with its decision to settle litigation on racial discrimination, helped control the damage to Texaco's image" (Brinson & Benoit, 1999, p. 507).

Even more clearly, the cases of Don Imus, Michael Richards, and Kelly Tilghman reveal an immediate decision toward the apology option. Tilghman, commentator for Golf Channel, responded to Nick Faldo's comment regarding Tiger Woods in a manner that created a ROC. Faldo was speaking of Woods' competitive dominance and stated the other players should "gang up" on him for a while. Tilghman replied with "lynch him in a back alley" (Sandomir, 2008). Tilghman apologized the next day. Likewise, Rush Limbaugh's assertion—that Donovan McNabb was receiving acclaim for his success on the football field because the National Football League wanted to promote black athletes—was met with a quick definitive response. In this case, Limbaugh resigned from the ESPN Sunday NFL Countdown show. While not in itself an apology, it was a response toward one extreme. There were no efforts to deny, diminish, or bolster in this case. The response would be more representative of a rebuilding posture designed to address the ROC created by the comment.

While fewer in number, there are cases in which an ROC response reflects a more assertive strategy. These strategies are found in the denial posture offered by Coombs (2007). The denial posture was present in response from a city and its mayor when they were accused of having roles in an ROC. In this case, the Hampton University women's basketball team traveled to Lubbock, Texas, to play the Texas Tech squad. After arriving in Lubbock, the coaches were detained and taken to the police station under suspicion of conducting a "pigeon drop" scam at a local Wal-Mart (Williams & Olaniran, 2002). It was eventually discovered that one of the coaches was incorrectly identified as the person attempting the fraud. However, this discovery came after several hours and the cancellation of the game. The incident quickly garnered national attention; at one point Johnny Cochran was retained to represent the Hampton coaches.

This ROC centered on the claims of the coaches that they were treated poorly and targeted because they are African American. The mayor of Lubbock argued the police department did nothing wrong. She claimed the Hampton coaches were detained because of a positive identification from the woman who was the target of the scam. The mayor and police chief also stated that the detainment and handling of the coaches were no different than that of any other person or group of people detained under those circumstances.

While an apology was eventually offered by the mayor, it took the form of an excuse strategy. The mayor, in addressing Hampton University, reiterated that the police officers were responding to a positive identification. The police chief stated race had nothing to do with the handling of the incident. At times when many expected apologies, the mayor and police chief instead distanced themselves, denying the wrong-doing, in the ROC.

The Hampton coaches eventually filed a lawsuit and were denied a final appeal in U.S. Circuit Court to have their lawsuit heard. In response to the decision, the mayor employed an attack-the-accuser strategy. She stated:

> I felt like our city was maligned maliciously by people who didn't live here. By people who were trying to make a name for themselves...I think ultimately this might have been about money, but I don't know that....We couldn't tell our side of the story, and by our side, I mean the truth. (Fuquay, 2001, p. 1).

This case represents a crisis response utilizing strategies (denial) from the other extreme. Denying the interpretation of the crisis event, the facts surrounding the crisis event, and then attacking the accuser, is a much more aggressive approach to responding to ROCs.

Anticipating ROCs

The anticipatory model advocated by Olaniran and Williams (2001) directs crisis management personnel to consider possible crisis-triggering events, within or related to an organization, within the internal and external foci. These specific considerations must be viewed in relationship to the organization's relational history, both internally and externally, with primary stakeholders. Therefore, crisis preparation (e.g., Coombs, 2007) and the identification of crisis vulnerabilities should include specific consideration of the possibility of (and history of) ROCs.

During crisis preparation and development of a crisis management plan (CMP), crisis management personnel should consider four steps:

1. Name an ROC review and response team (ROCRR) or individual.
2. Review the organization's relational history with stakeholders in terms of racial issues.
3. Evaluate potential internal triggers and external contributing factors.
4. Assess organizational readiness for an ROC.

ROCRR

The ROCRR or individual will be identified first. This person or group will then conduct the process described below. Selection of a single individual or a team of two to five individuals will depend upon size of the organization and the task of assessing organizational and industry history. The ROCRR team will need to include an individual or individuals with basic competencies in internal (organizational) and external (industry) research; rhetorical sensitivity; training program development, implementation, and evaluation; and a working knowledge of crisis management and response strategies. This responsibility (if held by one person) will likely be administered by someone from the public relations area. If an ROCRR team is developed, it should include at least one individual from public relations, and others can be drawn from areas relevant to tasks that need to be accomplished. The ROCRR should work with, or be a part of, those who develop the organizational crisis management plan. The work of the ROCRR will be significant at first, with a heavy research burden if that work has not already been done by those formulating the crisis management plan. After the ROCRR team has completed its initial work, the team will be able to shift to quarterly meetings provided there are no ROCs they need to address. In the event of an ROC, the ROCRR team will work with the crisis management team. The work of the ROCRR will begin with the research phase.

ROC Relational History

Reviewing organizational history in terms of race is consistent with suggestions signifying the importance of relational history and reputation management (e.g., Ledingham & Bruning, 2001; Ulmer, 2001). The crisis management team must look internally and externally to conduct an assessment of when, where, and how race has been part of a conflict or event in the organization. This assessment will identify what media and stakeholders will discover in the event of a new crisis episode involving race.

The assessment processes will need to include related organizations with a crisis history involving race. The attention drawn to a related organization following an ROC can give cause for media or stakeholders to call a similar ROC in one's own organization. For example, following Michael Richards's racist tirade during his comedy routine, attention was drawn to other comedians and entertainers who had similar incidents. With many of the subsequent incidents, media reports and public memory returned to the Richards incident, thus prolonging the effects of his own ROC. A similar public memory enactment happens when airlines and coal mines are faced with crises. A commercial plane crash or a coal mine collapse usually is reported and remembered in the context of those companies' individual safety records and a review of other airline and coal company safety records and history of similar crises.

Crisis management teams must recognize that their own racial incidents will generate public and stakeholder recognition of any previous ROCs they have managed. However, an ROC in a related organization might also create similar public and stakeholder recognition of their own ROC history. This process, while challenging, will be essential in identifying the organizations' crisis vulnerabilities (Fink, 1986; Pauchant & Mitroff, 1992).

Internal Triggers and External Factors

An evaluation of the potential internal triggers and external factors will follow the relational history assessment of the organization. This anticipatory step will allow review and response teams to identify areas where vulnerabilities for ROCs exist. The process will identify the potential for language and action triggers in the internal and external environments. Most ROCs will be triggered by either a language event or an action. Celebrity ROCs tend to result from a language trigger. Someone (e.g., Michael Richards, Don Imus, or Rush Limbaugh) says something that insults, unfairly critiques or criticizes, or demeans a race or member of that race, and an ROC has started. While the task is difficult, the ROC review and response teams must be vigilant in assessing the potential for language ROCs to develop.

Action-initiated ROCs develop when the organization implements a policy or practice that limits the safe and effective work or living conditions of its members or stakeholders. When such actions appear to include an element of race, the resulting crisis takes on the added dimension of an ROC. Oil companies that extract oil from regions of Nigeria have conducted practices that harm the Ogoni people of the Delta region. Past actions have polluted the environment and damaged the land and water, thus hindering the

Ogoni way of life and commerce. When word of these actions was transported out of the Delta region and concerned citizens across the globe became outraged, dominant oil companies faced an ROC stemming from their actions (Olaniran & Williams, 2008; Olaniran & Williams, in press).

Internal triggers would be those elements within the organization that could lead to an ROC, such as hiring and promotion practices, inter-organizational interactions, and workplace cultural norms and expectations. Texaco and the San Francisco 49ers crises are examples of ROCs that were triggered by internal factors. In these cases, a member (or members) of the organization who produced communication triggered an ROC. In the Texaco case, language was the initial trigger, which ignited the public reaction to the company. The San Francisco 49ers' ROC stemmed from the action of creating the demonstration video. Cases of improper hiring and promotion practices within an organization would be the most prevalent examples of actions leading to ROCs. For example, Shoney's struggled with an ROC in the 1990s stemming from the firing of African American members of its workforce (Watkins, 1997).

External factors would be those sources of influence outside the organization that could initiate (trigger), inflame, or deter an ROC. By definition, ROC review and response teams will have less control over these factors, and they can be unlimited in number. Still, the review process is essential to identifying the potential for an ROC. It may be difficult for the review and response team to identify each possible external contributing factor to an ROC, but efforts to identify potential areas of concern can be helpful in prevention and generation of a successful response strategy.

External triggers to ROCs are less frequent. The Hampton University and Lubbock, Texas, ROC was a case of an external triggering event that created a crisis for both the city of Lubbock and Texas Tech University. For Texas Tech specifically, the university became briefly involved in the ROC because the Hampton team had travelled to Lubbock to compete against the Lady Raiders, and the incident prevented the game from happening (Newton, 1998). Texas Tech became part of the ROC because the external triggering event and the external source (media) that reported the story were both connected to Texas Tech.

External factors (beyond triggering events) will typically include media, race-based organizations, shareholders, legal entities, and industry partners. The reaction from these (and other) entities will have influence on the duration and extent of damage resulting from the onset of an ROC.

Organizational Preparedness for ROCs

Assessing the organizational preparedness for an ROC will begin by following up on the results of the relational history, internal trigger, and external factor analyses. The ROCRR team will assess the organizational vulnerability to ROCs based upon the internal and external research. The team will rate the vulnerability on questions of:

1. the organization's history of internal and external ROCs and the potential of reoccurrence;
2. the industry history of internal and external ROCs and the potential of reoccurrence;
3. the primary triggers and contributing factors in previous organizational and industry ROCs;
4. the primary stakeholders and responding groups in previous organizational and industry ROCs; and
5. the current relational status of the organization with these stakeholders and groups.

Results of the relational history research will guide the ROCRR in reviewing and revising (if necessary) personnel training programs. Existing training programs may address the pertinent ROC issues already. If not, modifications in those programs, as needed, can be made to help in the anticipation and prevention of triggering events. Training sessions should highlight how words and actions can cause, or have caused, ROCs. Primary examples and case studies can come from organizational or industry examples. If those are not available or accessible, case studies can be drawn from available crisis communication literature (e.g., Fearn-Banks, 2001). To begin, training can take the shape of a communication effectiveness program (e.g., Dewine, 2001) with changes and insights into race-based issues and previous ROC cases.

Selection of ROC spokespersons is a key element in the review of personnel training. The organization likely has already identified individuals to serve as crisis spokespersons. Those individuals should receive first consideration for ROCRR spokespersons. If the normal crisis plan has not incorporated training for the crisis spokesperson, such training should be included for those who will possibly speak to the media and public about ROCs. Ulmer, Sellnow, and Seeger's (2007) guidelines for crisis spokespersons are a useful starting point for this training. Their suggestions for how to manage the media and demonstrate concern for those affected by the crisis as well as their general language strategies can be part of the guidelines provided for potential spokespersons. However, training scenarios should be developed and practiced, and spokespersons should be trained on their crisis response presentations.

Presentation manuscripts can be partially developed for spokespersons as the ROCRR team prepares response plans for potential crises. Findings for the relational history review will guide the ROCRR in terms of identifying potential responses and response strategies for the range of most likely ROCs. These preparations should be blended into the existing crisis response plan if the organization has one. All crisis responses will be contingent on the circumstances, and critical vigilant decision-making (e.g., Hirokawa & Rost, 1992; Williams & Olaniran, 1994) will be needed by the crisis response team, ROCRR, and primary organizational decision-makers when a crisis occurs. This decision-making will be enhanced with the preparations conducted by the ROCRR. The ROCRR will be able to establish parameters for the type of response that would be feasible given the crisis event and what the likely reactions will be from media, stakeholders, and the general public based on the relational history research. In essence, decision-makers will be able to make better informed decisions in a timely manner.

Conclusion

The onset of an ROC can be devastating to an organization or individual. These crises tend to garner quick and far-reaching media and stakeholder attention. The inception of an ROC also tends to trigger the memory of any previous ROC in the organization or even in a related organization. The quick and long-term damage that can afflict an organization or individual caught in a ROC requires diligence in the anticipation and prevention of such events. Recognizing its importance is the initial step in the anticipation and preparation for ROCs. The guidelines offered above should assist organizations in the development of procedures to limit the likelihood of an ROC or, more effectively, manage the situation if an ROC does develop.

References

Adamson, J. (1998, October 5). The Denny's discrimination story—and ways to avoid it in your operation. *Nations Restaurant Review, 32*, 40.

Baker, G. F. (2001). Race and reputation: Restoring image beyond the crisis. In R. L Heath (Ed.), *Handbook of public relations* (pp. 513–520). Thousand Oaks: Sage.

Brinson, S. L., & Benoit, W. L. (1999). The tarnished star: Restoring Texaco's damaged public image. *Management Communication Quarterly*, *12*, 483–510.

Carr, D. (2007, April 7). Networks condemn remarks by Imus. *The New York Times*. Retrieved January 26, 2009, from http://www.nytimes.com/2007/arts/television/07/imus.html

Chin, T., Naidu, S., Ringel, J., & Snipes, W. (1998). Denny's communication amidst a discrimination case. *Business Communication Quarterly, 61*, 180–197.

Coombs, T. (2007). *Ongoing crisis communication.* (2nd ed.). Los Angeles: Sage.

Dewine, S. (2001). *The consultant's craft* (2nd ed.). Boston: Bedford/St. Martin's.

Fearn-Banks, K. (2001). *Crisis communications: A casebook approach* (2nd ed.). Mahwah, NJ: Lawrence Erlbaum.

Fink, S. (1986). *Crisis management: Planning for the inevitable*. New York: AMACOM.

Fuquay, J. (2001, January 11). Hampton suit draws Sitton fire. *Lubbock Avalanche-Journal*, p. 1.

Hirokawa, R. Y., & Rost, K. (1992). Effective group decision making in organizations. *Management Communication Quarterly, 5*, 267–288.

Ledingham, J. A., & Bruning, S. D. (2001). *Public relations as relational management.* Hillsdale, NJ: LEA.

Newton, C. (1998, November 18). Hampton loses chance to play No. 12 Texas Tech. *Associated Press*. http://foxnews.com/sports/wires2/1118/s_ap_1118_41.sml

Olaniran, B. A., & Williams, D. E. (2001). Anticipatory model of crisis management: A vigilant response to technological crises. In Heath, R. L. (Ed.), *Handbook of public relations* (pp. 487–500). Thousand Oaks: Sage.

———. (2008). Applying anticipatory and relational perspectives to the Nigerian Delta Region oil crisis. *Public Relations Review, 34,* 57–59.

———. (2005). Protecting the environment and people's well-being in the Nigerian Ogoni land. In M. Parkinson & D. Ekachai (Eds), *International and intercultural public relations: A campaign case approach* (320–332). Boston: Allyn and Bacon.

Pauchant, T. C., & Mitroff, I. I. (1992). *Transforming the crisis-prone organization: Preventing individual, organizational, and environmental tragedies.* San Francisco: Jossey-Bass.

Racism is to be expected from Don Imus. (2007, April 9). *Fairness & accuracy in reporting*. Retrieved November 11, 2007, from www.fair.org/index.php?page=3082

Richards hires "crisis management" expert. (2007, November 22). *MSNBC*. Retrieved from http://www.msnbc.msn.com/id/15859121/http://www.theolympian.com/2010/06/16/1275880/bp-agrees-to-set-aside-20-billion.html

Sandomir, R. (2008, January 11). When an apology is not enough. *The New York Times*. Section D, p. 3.

Ulmer, R. R. (2001). Effective crisis management through established stakeholder relationships. *Management Communication Quarterly, 14*, 590–615.

Ulmer, R. R., Sellnow, T. L., & Seeger, M. W. (2007). *Effective crisis communication: Moving from crisis to opportunity*. Thousand Oaks: Sage.

Watkins, S. (1997). *The black O: Racism and redemption in an American corporate empire.* Athens: Georgia UP.

Williams, D. E., & Olaniran, B. A. (1994). Exxon's decision-making flaws: The hypervigilant response to the Valdez grounding. *Public Relations Review*, *20*(1), 5–18.

———. (2002). Crisis communication in racial issues, *Journal of Applied Communication Research, 30*, 293–313.

Conclusion

Bolanle A. Olaniran & David E. Williams

The title *Pre-crisis Planning, Communication, and Management: Preparing for the Inevitable* offers a significant insight into the critical considerations of crisis preparation. However, this book also signifies the growth in understanding and appreciation for the complexities of crisis management as a whole.

Early thoughts about crisis management were devoted to a management-oriented prescriptive approach to dealing with business catastrophes. In the 1970s and 1980s manuals were prepared, and some books were written to help guide public relations executives and mid-level managers through the rigors of dealing with what came to be known as "the inevitable." Somewhat simultaneously, the costs of crises and recognition of the importance of managing crises grew through the 1980s and into the 1990s.

During that period of growth, a critical eye developed among crisis management and crisis communication scholars. This early attention frequently focused on post-event analyses of crises, most typically through a rhetorical lens. Many of these insightful works would examine in detail the words and actions taken by organizations under crises. The spokesperson's communication and organization's actions would be viewed with a somewhat Neo-Aristotelian perspective in an effort to determine why stakeholder and public reaction took whatever form it took. These types of critical insights quickly led to more theoretical perspectives and applications from other areas of communication, risk management, business, and a host of other disciplines.

With the turn of the century we had models (or more models) of crisis lifecycles, which showed the stages an organization will typically traverse when responding to a crisis. We also came to understand the various crisis types as well as the differences between them. This research did not stop at post hoc analyses of what different organizations did well or poorly in their crisis communication efforts. It also produced theories on how to respond to specific types of crises and parts of the crisis lifecycle. This research lineage leads to where we are now with the title *Pre-crisis Planning, Communication, and Management: Preparing for the Inevitable*. The anticipatory perspective represents not the last but the most recent stage of intellectual growth in crisis management research. While pre-crisis planning has been a part of crisis management since practitioners and scholars started devoting attention to crises, the insights have never been as detailed and thorough as those represented in this work.

Thirty years ago, crisis management had very little theoretical grounding, with few people suggesting the need for such effort. Most were content to describe best practice scenarios or offer insights from personal experience working in crisis management. Now, even the pre-crisis stage of crisis management is discussed in both applied and theoretical terms. This work offers the anticipatory perspective as a means for guiding those discussions, but it is not the only insight. The initial section of *Pre-crisis Planning, Communication, and Management: Preparing for the Inevitable* is certainly representative of the work done in theory and model development in crisis preparation research. Communication theory is replete with mechanisms for refining the understanding of our environment and projecting likely outcomes given specified variables. It is a logical sign of intellectual growth when we witness crisis management and crisis communication traveling the same path of environmental interpretation and understanding. We would anticipate this to be an area of tremendous growth in the future.

Practical necessity among businesses and organizations combined with the natural growth of theory development should spur a great deal of theoretical research on pre-crisis management in the future. Public relations experts, and crisis management personnel, in particular, have done an excellent job over the years of drawing attention to the grand scope of risks associated with crises. Even in challenging economic conditions, organizations are aware of the costs of crises and are willing to put forth effort and resources in crisis prevention. Extensive use of training programs is one example of this desire to avoid crisis situations. Simply put, there is a market need for continued work in pre-crisis planning, and crisis scholars have an excellent body of research to develop further. The working relationship between crisis communication and risk communication only strengthens that need for more research in crisis preparation.

There is no doubt the events on September 11, 2001, had a significant impact on crisis management research. That day clearly demonstrated a relationship between crisis and risk that scholars and practitioners had already noticed. The second section of this work recognizes that relationship as well. Terrorist attacks continue to be the initial consideration when we think of the interdependency of crisis preparation and response, emergency management, and risk detection. However, we have seen how natural disasters, domestic attacks, and even individual misdeeds can call into action responses from individuals in all of these areas. It is reasonable and necessary that crisis management and communication scholars work closely with those in these related areas, as the crises of the day often require expertise in all areas. Anticipation of crises, therefore, must take into consideration the expertise of individuals working in these areas.

Section three of this work provides assurance that the reader recognizes the challenge of preparing for a crisis in today's environment. Our research in the 1980s tended to assume an American organization, working with an American audience and American stakeholders. This allowed for more of a one-approach, or one-response strategy, fits all. The multicultural nature of today's organization necessitates recognition of cultural, ethnic, and racial differences. Multicultural differences will alter how a crisis is interpreted, how communication during the crisis will be interpreted, and how an organization or individual will be viewed during and after the crisis. Therefore, it is paramount that crisis preparation efforts fully recognize the range of individuals who will have a role in a potential crisis.

Our final section addresses the ethical nature inherent in crisis communication. Ethics, or ethical violations, are often the source of organizational or individual crises. Ethical crises can be terribly challenging and damaging as they impugn an organization in a manner that is hard to reverse or forgive. Because of these extreme dangers from ethical violations, the responses to these crises are often quick and dramatic in an effort to rid the organization of the misconduct. However, ethical considerations in crisis management extend beyond the response to such incidents and into the practice of crisis management itself.

As with all communication endeavors, crisis management practitioners must be ever vigilant of the ethical integrity of their actions. Crisis communication is typically scrutinized by those involved in the crisis. Stakeholders, media, individuals affected by the crisis, and frequently even the general public in cases of significant crisis all attend very closely to what an organization does and says in response to a crisis. This is no time for an ethically suspect action or communication. Those in charge of pre-crisis preparation efforts must not lack in their understanding of the importance of ethical communication practices.

Pre-crisis Planning, Communication, and Management: Preparing for the Inevitable stands as an example of how crisis management and crisis communication have grown as a field in the last 30 years. We hope this work can also serve as an example of where future research can go. The anticipatory perspective is one approach for examining one element of crisis management. We anticipate the future of crisis management and crisis communication will include similar theory-driven, detailed examinations of the place of crisis management and crisis communication in the world.

Contributors

Shannon A. Bowen (PhD, University of Maryland) is associate professor in the Newhouse School. Bowen specializes in public relations ethics, strategic communication management, issues management, crisis communication, dominant coalition decisions, and organizational communication. Bowen is joint editor of *Ethical Space: The International Journal of Communication Ethics.* Bowen has also won numerous top paper awards and has been published in leading journals of the field. In 2010, she published the book *An Overview of the Public Relations Function* with Brad Rawlins and Thomas Martin as the first volume in the public relations series from Business Expert Press. In 2011, she won the Jackson Jackson and Wagner Behavioral Science Research Prize.

Glen T. Cameron (PhD, University of Texas—Austin) is the Maxine Wilson Gregory chair in journalism research and professor of family and community medicine at the University of Missouri. He directs the Health Communication Research Center at the University of Missouri. In bibliometric analyses, Cameron is cited as the most published researcher nationally in major refereed journals over the past twenty years. Recipient of the Pathfinder Award for Career Research Contributions from the Institute of Public Relations Research & Education, Cameron's research includes studies of public relations, strategic health communication, and conflict management. He is co-author of a leading introductory text, *Public Relations: Strategies and Tactics*, now in its tenth edition and available in ten languages.

Robert C. Chandler (PhD, University of Kansas) is professor of communication and director of the Nicholson School of Communication at the University of Central Florida. Chandler researches organizational integrity, ethics training and assessment, business continuity, and business ethics. He is a national board member of the Center for the Public Trust as well as for e-businessethics.com. He has authored more than 75 academic and professional publications including his co-authored book *Managing the*

Risks for Corporate Integrity: How to Survive an Ethical Misconduct Disaster. He also investigates organizational communication, crisis communication, emergency notification, message comprehension, conflict management, risk communication, multicultural, and intercultural communication.

Venetia Tzi-Wei Chung (Bachelor of Communication Studies, Singapore Nanyang Technology University's Wee Kim Wee School of Communications and Information) has developed a keen interest in crisis management and communication during an internship in a public relations agency, Asia PR Werkz. In her final year undergraduate project, Chung was involved in the planning and execution of a campaign that saw the launch of the Singapore 2010 Youth Olympic Games Logo. After she graduated, Chung joined Marina Bay Sands, an integrated resort in Singapore as a corporate communications officer. In her job, she manages both global and regional media and assisted in the organization's crisis planning.

Jesper Falkheimer *(*PhD, Lund University) is associate professor in strategic communication at Lund University in Sweden and rector at Campus Helsingborg, Lund University, Sweden. He has published several books, chapters in international anthologies, and articles in journals such as *Public Relations Review*, *International Journal of Strategic Communication,* and *Journal of Crisis and Contingencies Management*. His research is founded in social theory as well as strategic communication theory, and he has done studies in areas such as crisis communication, place branding, news management, and organizational communication.

Karen Freberg (PhD, University of Tennessee) is an assistant professor in strategic communication at the University of Louisville. She is also an adjunct instructor for the Integrated Marketing Communications (IMC) Graduate Online Program at West Virginia University. Freberg's research interests are in public relations, social media, crisis communications, and reputation management. Freberg has presented at several U.S. and international research conferences, including ones in the Netherlands, Brazil, Greece, and China.

Robert Heath (PhD, University of Illinois) is professor emeritus at the University of Houston, author of many books and articles on issues management, organizational rhetoric, public relations, crisis communication, and risk communication.

Mats Heide (PhD, Lund University) is associate professor in strategic communication at Lund University in Sweden. He is responsible for the master's program in strategic communication at the Department of Communication and Media at Lund University. Heide has published eight books (in Swedish), some chapters in international anthologies, and articles in journals such as *Corporate Communications*, *Nordicom Review*, and *Journal of Crisis and Contingencies Management.* His research interests are in strategic communication, and he has mainly focused on organizational communication. From 2011 through 2013, Heide will work with a research project on internal crisis communication, which encompasses and will combine crisis communication and change communication.

Tony Jaques (PhD, RMIT University, Melbourne, Australia) is managing director of issue outcomes in Australia and specializes in issue and crisis management and risk communication. He was formerly Asia-Pacific Issue Manager for an American corporation and has been widely published in journals including *Journal of Communication Management, Journal of Public Affairs, Public Relations Quarterly, Journal of Business Strategy, Public Relations Review, Corporate Communication, Journal of Public Relations, Disaster Prevention and Management*, and *International Journal of Strategic Communication.* He also teaches in the master's program at RMIT University in Melbourne, Australia, and writes the online issue and crisis newsletter *Managing Outcomes* (www.issueoutcomes.com.au).

Yan Jin (PhD, University of Missouri-Columbia) is an associate professor at the School of Mass Communications, Virginia Commonwealth University. Her research focuses on crisis communication, strategic conflict management, and how emotions influence public relations decision-making and publics' responses. She has published widely in refereed journals and textbooks and has won top paper awards at international conferences.

Robert S. Littlefield (PhD, University of Minnesota, 1983) is a professor of communication and director of the Risk+Crisis Communication Project at North Dakota State University in Fargo. His research interests include risk and crisis communication, intercultural communication, and emergency management. He is co-author of *Effective Risk Communication: A Message-Centered Approach,* and has published in *Journal of Applied Communication Research, Journal of Intercultural Communication Research, Risk Analysis, Journal of Emergency Management, Communication Culture & Critique,* and *International Journal of Business Continuity and Risk Management.*

Luana Sok-Yee Low (Bachelor of Communication Studies, Singapore Nanyang Technology University's Wee Kim Wee School of Communications and Information) developed a keen interest in the area of crisis communications during her internship at global public relations firm Burston Marsteller. Low is currently with the global public relations agency Fleishman-Hillard, where she assists her team on various major accounts in the consumer and technology practices. In her final year undergraduate project, Low was involved in the planning and execution of a campaign that saw the launch of the Singapore 2010 Youth Olympic Games Logo. After she graduated, Low was an assistant corporate communications manager with the Civil Aviation Authority of Singapore, where she gained experience in international relations and crisis management.

Clara Mitcham (BA, University of Oklahoma) is a national leadership consultant for Alpha Gamma Delta. Her honors research program focused on under-represented populations in crises.

Lan Ni (PhD, University of Maryland—College Park) is an assistant professor in the Jack J. Valenti School of Communication at the University of Houston. Her research focuses on strategic management of public relations, identification of publics, relationship management, intercultural public relations, and internal communication. She has published in major journals such as *Journal of Public Relations Research*, *Public Relations Review*, *Journalism and Mass Communication Quarterly*, *Journal of Communication Management*, and *International Journal of Strategic Communication*. She currently serves on the editorial board of *Journal of Public Relations Research* and the Research Advisory Committee of the International Association of Business Communicators (IABC) Research Foundation. She has presented more than 20 papers at conferences such as ICA, AEJMC, NCA, PRSA, and International Public Relations Research Conference.

Bolanle A. Olaniran (PhD, University of Oklahoma) is a professor in the Department of Communication Studies at Texas Tech University. His research includes crisis communication, organization communication, cross-cultural communication, and technologies. He has authored several peer reviewed articles in discipline focus and interdisciplinary focus journals (i.e., regional, national, and international) and several edited book chapters in each of these areas. He has served as a consultant to organizations and universities at local, national, international and government levels. His works have gained recognition, such as the American Communication Association's Outstanding Scholarship in Communication Field, *Who's Who among College Teachers*, *Who's Who in Sciences and Engineering*, and others. He is also the recipient of several awards including the 2008 President's Excellence Teaching Award and the 2006 Office of the President's Diversity Awards. Recent international invited lecture series include countries like Taiwan, Russia, and Hong Kong,

Michael J. Palenchar (PhD, University of Florida) is associate professor and co-director of the Risk, Health and Crisis Communication Research Unit at the University of Tennessee. Palenchar has more than two decades of professional and academic public relations experience. His research interests include risk and crisis communication, issues management, community right-to-know issues, front groups, the use of social media during a crisis, and general public relations. He is a risk communication and issues management research consultant for clients ranging from Fortune 500 companies to local government and nongovernmental agencies. His research has been published, for example, in *Management of Communication Quarterly, Journal of Public Relations Research, Public Relations Review, Public Relations Journal, Environmental Communication, Journal of Contingencies and Crisis Management*, and among other sources. His first book with co-author Robert L. Heath entitled *Strategic Issues Management* (2nd edition) was published by Sage in 2009.

Augustine Pang (PhD, University of Missouri—Columbia) is an assistant professor at the Singapore Nanyang Technology University's Wee Kim Wee School of Communications and Information. Pang's research interests include crisis management and communication, image management and repair, public relations, and media relations. He has been published widely in refereed journals and textbooks and has won top paper awards at international conferences.

Timothy L. Sellnow (PhD, Wayne State Univerity) is Graduate Programs associate dean and professor of communication in the College of Communications and Information Studies at the University of Kentucky. His research focuses on bioterrorism, pre-crisis planning, and instructional communication strategies for mitigating personal harm during crises. He has conducted funded research for the Department of Homeland Security, the U.S. Department of Agriculture, and the Centers for Disease Control and Prevention.

Judy Sultan (MA, University of Houston) continues research and professional practice to advance sports journalism, sports marketing, and sound sports management. She published her thesis, *Narrative Analysis of Crisis: "Beware of Dog" or Be Blitzed by the Vick-Timized Voice* under the direction of Robert Heath in 2009.

Jennifer Vardeman-Winter (PhD, University of Maryland—College Park) is an assistant professor at the University of Houston's Jack J. Valenti School of Communication. She is also an affiliate faculty member with University of Houston's Women's Studies Program. Her studies are concentrated in public relations campaigns, health communication, and multicultural feminist research. Her work has been published in *Journal of Public Relations Research, Public Relations Review*, *Health Communication*, and *Feminist Media Studies*. She also recently worked for Macro International, where she consulted for governmental health agencies such as the CDC's Radiological Studies Branch, the Department of Homeland Security's Citizen Corps, and the National Cancer Institute's Survey of Health and Medical Science Reporters and Editors. She currently consults for Baylor College of Medicine's Teen Health Clinics in developing theory-based Webisodes to increase safe sex practices among African American and Latino low-income teens.

Shari R. Veil (PhD, North Dakota State University) is director of risk sciences and assistant professor of communication in the College of Communications and Information Studies at the University of Kentucky. Her research focuses on organizational learning in high-risk environments, community preparedness, and communication strategies for crisis management. Veil's work has been funded by the U.S. Department of Agriculture, the Department of Homeland Security, and Environmental Protection Agency. She has been published in venues such as *Journal of Applied Communication Research, Management Communication Quarterly, Journal of Business Ethics, Journal of Business Communication,* and *Public Relations Review.*

Damion Waymer (PhD, Purdue University) is an assistant professor of communication at Virginia Tech. His research uses issues management, public relations theory, and organizational rhetoric to shed light on issues of culture and diversity in general and issues of race, class, and gender specifically. Additionally, his research uses communication theories to explore some of the issues that marginalized or under-represented publics can and do encounter and what strategies are available to them to challenge these issues. Representative works have been published in outlets such as the *Journal of Applied Communication Research*, *Public Relations Review*, *Communication Quarterly*, and the *Journal of Communication Inquiry*.

David E. Williams (PhD, Ohio University) is a retired faculty member of the Department of Communication Studies at Texas Tech University. Williams taught at Texas Tech for 18 years until 2009 and currently resides in Wildwood, Missouri. Williams' work in crisis communication has been published in journals such as the *Journal of Applied Communication Research*, *Public Relations Review*, and *Communication Studies* as well as several book chapters. Williams continues to work in crisis communication research and communication skills development and training.

Index

A

B

C

D

E

I

J

K

L

M

N

O

P

Q

R

S

T

U

V

W

Y

Z